Overconvergence in Complex Approximation

Sorin G. Gal

Overconvergence in Complex Approximation

Springer

Sorin G. Gal
Department of Mathematics
  and Computer Science
University of Oradea
Oradea, Romania

ISBN 978-1-4899-9791-3   ISBN 978-1-4614-7098-4 (eBook)
DOI 10.1007/978-1-4614-7098-4
Springer New York Heidelberg Dordrecht London

Mathematics Subject Classification (2010): 30E10, 30G35, 33C45, 41A35, 41A28, 41A25

To my grandsons Tudor and Darius

Preface

It is a well-known fact that the concept of overconvergence in approximation theory may have several meanings. The most common, developed for the first time by Ostrovski and Walsh, is that given a sequence of functions approximating a given (analytic) function in a set (region), the convergence may hold not merely in that set, but in a larger one containing the first set in its interior.

A second meaning is the well-known Walsh's overconvergence phenomenon in interpolation of functions introduced by Walsh in [144], p. 153, intensively studied by many mathematicians (see, e.g., the recent research monograph by Jakimovski–Sharma–Szabados [88]), which briefly can be described as follows: given $r > 1$ and the function $f(z) = \sum_{k=0}^{\infty} a_k z^k$, $|z| < r$, denoting by $L_{n-1}(f)(z)$ the Lagrange polynomial of degree $\leq n-1$ interpolating f in the n-th roots of unity, we have

$$\lim_{n \to \infty} \left(L_{n-1}(f)(z) - \sum_{k=0}^{n-1} a_k z^k \right) = 0, \text{ for all } z \in \mathbb{C} \text{ with } |z| < r^2,$$

the convergence being uniform and geometric in any compact subset of $\{z \in \mathbb{C}; |z| < r^2\}$. Moreover, the result is best possible, that is, it does not hold at any point $|z| = r^2$.

The third meaning refers to the overconvergence of power series $f(z) = \sum_{j=0}^{\infty} a_j z^j$ with the radius of convergence $R \geq 0$ and it can be described as follows (see, e.g., Bourion [22], Ilieff [84], Walsh [143], Luh [98], Kovacheva [92], Beise–Meyrath–Müller [14]): denoting $S_n(f, z) = \sum_{j=0}^{n} a_j z^j$, $n \in \mathbb{N}$, if there exist a subsequence $(S_{n_k})_{k \in \mathbb{N}}$ and a domain U containing the open disk $\mathbb{D}_R$ as a proper set, such that $S_{n_k}(f, z)$ converges inside of U to an $S \in \mathbb{C}$ as $k \to \infty$, then the power series is called overconvergent.

The overconvergence studied in this research monograph belongs to the first meaning of the phenomenon and consists in two directions, which briefly can be described as follows:

1) Let $I \subset \mathbb{R}$ be a subinterval, $C(I) = \{f : I \to \mathbb{R}; f$ is continuous on $I\}$ and $L_n : C(I) \to C(I)$, $n \in \mathbb{N}$, a given sequence of approximation operators with the property that for any $f \in C(I)$ we have $\lim_{n\to\infty} L_n(f)(x) = f(x)$ for all $x \in I$ (pointwise or uniformly).

We say that the overconvergence phenomenon holds for the sequence $(L_n)_n$ if there exists $G \subset \mathbb{C}$ containing I (e.g., if I is a compact real interval, then G might be a compact disk containing I), such that for any function $f : G \to \mathbb{C}$ analytic on G we have $\lim_{n\to\infty} L_n(f)(z) = f(z)$, for all $z \in G$ (pointwise or uniformly).

Note that instead of the space $C(I)$, more generally we can consider the space $C(I; X) = \{f : I \to X; f$ is continuous on $I\}$, where $(X, \|\cdot\|)$ is a complex Banach space.

2) For $f : \mathbb{R} \to \mathbb{R}$ and the kernel $K : \mathbb{R} \times \mathbb{R}_+ \to \mathbb{R}$, let us consider the integral convolution operator $C_t(f)(x) = \int_{-\infty}^{+\infty} K(u,t)f(x-u)du$, $t \geq 0$, with the property that for any f in a subclass of continuous functions, we have $\lim_{t\to 0} C_t(f)(x) = f(x)$, for all $x \in \mathbb{R}$ (pointwise or uniformly). In this case, besides the overconvergence phenomenon defined at point 1, we can consider another one, as follows.

We say that for $(C_t(f))_{t\geq 0}$, the *convolution overconvergence phenomenon* holds if by replacing in the formula of the integral convolution $C_t(f)(x)$ the translation $x - u$ with the rotation ze^{iu}, then for any analytic function f in a compact disk or subset $G \subset \mathbb{C}$, we have $\lim_{t\to 0} C_t^*(f)(z) = f(z)$, for all $z \in G$ (pointwise or uniformly), where at this time $C_t^*(f)(z) = \int_{-\infty}^{+\infty} K(u,t)f(ze^{iu})du$.

Evidently, besides the qualitative aspects just mentioned above, the overconvergence phenomenon presents quantitative aspects too regarding the order of approximation of f by $L_n(f)$ or by $C_t^*(f)$ on G. An important characteristic of all these results (which does not hold in the case of real approximation) is that the approximation orders obtained are exact.

The history of the overconvergence phenomenon in complex approximation by Bernstein-type operators goes back to the work of Wright [146], Kantorovich [89], Bernstein [16–18], Lorentz [96] (Chap. 4), and Tonne [139], who in the case of complex Bernstein operators defined by

$$B_n(f)(z) = \sum_{k=0}^{n} p_{n,k}(z)f\left(\frac{k}{n}\right), \quad p_{n,k}(z) = \binom{n}{k}z^k(1-z)^{n-k}, |z| \leq r,$$

have given interesting qualitative results, but without giving quantitative estimates. Also, qualitative results without any quantitative estimates were obtained for the complex Favard–Szász–Mirakjan operators by Dressel–Gergen and Purcell [32] and for the complex Jakimovski–Leviatan operators by Wood [147]. We notice that the qualitative results are theoretically based on the "bridge" made by the classical result of Vitali (see Theorem 1.1.1), between the (well-established) approximation results for these Bernstein-type operators of real variable and those for the Bernstein-type operators of complex variable.

In the very recent book of Gal [49], a systematic study of the overconvergence phenomenon in complex approximation was made for the following important classes of Bernstein-type operators: Bernstein, Bernstein–Faber, Bernstein–Butzer, q-Bernstein with $0 < q \leq 1$, Bernstein–Stancu, Bernstein–Kantorovich, Favard–Szász–Mirakjan, Baskakov, and Balázs–Szabados.

Also, in the same book, the convolution overconvergence phenomenon in the sense of the above direction 2) was studied, for the following types of integral convolution operators: de la Vallée Poussin, Fejér, Riesz–Zygmund, Jackson, Rogosinski, Picard, Poison–Cauchy, Gauss–Weierstrass, q-Picard, q-Gauss–Weierstrass, Post–Widder, rotation invariant, Sikkema, and nonlinear.

The aim of this book is to continue these studies, naturally completing and generalizing the results in the previously mentioned book, as follows.

In the sense of the above-mentioned direction 1), we present here similar results for the Schurer–Faber operator, Beta operators of the first kind, Bernstein–Durrmeyer-type operators and Lorentz operator.

But unlike the previous book of Gal [49], here in six sections (Sects. 1.8–1.13) we also consider the approximation by several complex q-Bernstein-kind operators for $q > 1$, the case when they give the geometric order of approximation $O(1/q^n)$ (which is nearly to the best approximation). It is worth noting that the q-approximation problem with $q > 1$ is considered here not only in compact disks of $\mathbb{C}$ for various q-approximation operators but also in compact disks of the noncommutative field of quaternions for the q-Bernstein operator (see Sect. 1.12) and in compact subsets of the complex plane $\mathbb{C}$ for the so-called q-Bernstein–Faber polynomials (see Sect. 1.11) and q-Stancu–Faber polynomials (see Sect. 1.9). We emphasize that the results in Sect. 1.11 represent natural and strong extensions of the approximation results for the q-Bernstein polynomials in $[0,1]$ to various compact subsets of the complex plane, for example, the compact disks, the circular lunes, the annulus sectors, the compact set bounded by the m-cusped hypocycloid H_m, $m = 2, 3, \ldots$, given by the parametric equation

$$z = e^{i\theta} + \frac{1}{m-1}e^{-(m-1)i\theta}, \ \theta \in [0, 2\pi),$$

the regular m-star $(m = 2, 3, \ldots,)$ given by

$$S_m = \{x\omega^k; 0 \leq x \leq 4^{1/m}, k = 0, 1, \ldots, m-1, \omega^m = 1\},$$

the m-leafed symmetric lemniscate, $m = 2, 3, \ldots$, with its boundary given by

$$L_m = \{z \in \mathbb{C}; |z^m - 1| = 1\},$$

and the semidisk

$$SD = \{z \in \mathbb{C}; |z| \leq 1 \text{ and } |Arg(z)| \leq \pi/2\}.$$

The approximation results obtained refer to exact estimates in approximation and in simultaneous approximation and to quantitative Voronovskaja-type results.

In the sense of the above-mentioned direction 2), quantitative overconvergence and convolution overconvergence results are presented here for the convolution potentials generated by the Beta and Gamma Euler's functions.

Finally, the overconvergence phenomenon in the sense of direction 1) for the most classical orthogonal expansions (of Chebyshev, Legendre, Hermite, Laguerre, and Gegenbauer kinds) attached to vector-valued functions is studied.

More detailed, the book can be described as follows.

The structure of Chap. 1 is the following:

- Section 1.1 contains the main results and concepts in complex analysis required for the proofs of the results in this book. For example, we mention here the following: the Vitali's theorem, Cauchy's formula, Bernstein's inequality, Faber polynomials associated with a domain in $\mathbb{C}$, Faber series, and Faber coefficients.
- In Sect. 1.2 the exact order in the generalized Voronovskaja's result for the derivatives of the complex Bernstein polynomials is obtained, thus generalizing the Voronovskaja's theorem for the Bernstein polynomials in Gal [49], pp. 36–42.
- In Sects. 1.3–1.7 we prove similar properties with those obtained for the complex Bernstein polynomials in Gal [49], Chap. 1, for the following classes of complex operators: Schurer–Faber polynomials, Beta operators of the first kind, genuine Bernstein–Durrmeyer polynomials, Bernstein–Durrmeyer polynomials with Jacobi weights, and Lorentz polynomials, respectively.
- In Sects. 1.8–1.12, error estimates of order $\frac{1}{q^n}$, $q > 1$, in approximation by complex q-Lorentz polynomials, q-Stancu and q-Stancu–Faber polynomials, q-Favard–Szász–Mirakjan operators, q-Bernstein–Faber polynomials, and q-Bernstein polynomials of quaternion variable, respectively, are obtained.
- Section 1.13 contains notes and open problems including the approximation by q-Lorentz–Faber, q-Bernstein–Kantorovich, q-Szàsz–Kantorovich, and q-Durrmeyer polynomials, with $q > 1$.

In Chap. 2 we present the overconvergence phenomenon in strips and the convolution overconvergence of the integral convolutions with trigonometric kernels including the Beatson kernel and its iterates and the approximation by complex potentials generated by the Euler-type functions.

Chapter 3 studies the overconvergence phenomenon in the sense of direction 1) with explicit quantitative estimates of geometric order, for the orthogonal expansions of Chebyshev and Legendre kinds attached to vector-valued functions. It is worth noting that in the last Sect. 3.4, one presents with details some interesting open problems concerning the possible application of the results obtained to the orthogonal systems of Hermite polynomials, Laguerre polynomials, and Gegenbauer polynomials.

Let us mention that most of the results presented here have been obtained by the author of this monograph in a series of papers, single or jointly written as can be seen in the bibliography. Also Theorems 1.2.7, 1.3.1, 1.7.7, 1.11.4, 2.1.1, 2.2.3, 2.2.4, 2.2.5, 3.2.4, 3.3.1, and 3.3.3 and Corollaries 1.11.5 and 3.2.5 appear for the first time in this book.

It is worth noting that this book suggests for future research, similar studies for other complex linear and nonlinear convolutions and for other Bernstein-type operators (including approximation in compact disks in $\mathbb{C}$, in compact disks in the field of quaternions and in compact subsets in $\mathbb{C}$ by their Faber-kind variants), like those of Meyer–König–Zeller type, Jakimovski–Leviatan type, Bleimann–Butzer–Hahn type, and Gamma type (including their q-variants with $q > 0$). For other examples of approximation operators to which the overconvergence theory could be applied, see Sect. 1.13.

The book mainly is addressed to researchers in the fields of the complex approximation of functions and its applications, mathematical analysis, and numerical analysis.

Also, since most of the proofs use elementary complex analysis, it is accessible to graduate students and suitable for graduate courses in the above domains.

I would like to thank Vaishali Damle, senior editor at Springer publisher and Ethiraju Saraswathi, project manager at SPi Global, for their great cooperation.

Oradea, Romania   Sorin G. Gal

Contents

Chapter 1

Overconvergence in $\mathbb{C}$ of Some Bernstein-Type Operators

Section 1.1 of this chapter contains classical definitions and results in complex analysis useful for the next sections.

In Sect. 1.2 the exact order in the generalized Voronovskaja's result for the derivatives of the complex Bernstein polynomials is obtained, thus generalizing the generalized Voronovskaja's theorem for the complex Bernstein polynomials in Gal [49], pp. 36–42.

In Sects. 1.3–1.12 we prove similar properties with those obtained for the complex Bernstein operators studied in Gal [49], Chap. 1, for the following classes of complex operators: Schurer–Faber, Beta of the first kind, genuine Bernstein–Durrmeyer, Bernstein–Durrmeyer with Jacobi weights, Lorentz, q-Lorentz with $q > 1$, q-Stancu and q-Stancu–Faber with $q > 1$, q-Favard–Szász–Mirakjan with $q > 1$, q-Bernstein–Faber with $q \geq 1$, and q-Bernstein of quaternion variable with $q \geq 1$, respectively. For all these Bernstein-type operators, the exact degrees of approximation are obtained by three steps: (1) upper estimates, (2) quantitative Voronovskaja-type formula, and (3) lower estimates by using the step 2.

1.1 Auxiliary Results in Complex Analysis

In order to make the book more self-contained, in this section we briefly present the main classical results in complex analysis that will be used in the book.

The first one, called Vitali's theorem, can be stated as follows.

Theorem 1.1.1 (Vitali; see, e.g., Kohr–Mocanu [91], p. 112, Theorem 3.2.10). *Let Ω be a domain in $\mathbb{C}$ and $F \subset \Omega$ a set having at least one accumulation point in Ω. If the sequence of analytic functions in Ω, $(f_n)_{n \in \mathbb{N}}$, is bounded in each compact in Ω and $(f_n(z))_n$ is convergent for any $z \in F$, then $(f_n)_{n \in \mathbb{N}}$ is uniformly convergent in any compact subset of Ω.*

S.G. Gal, *Overconvergence in Complex Approximation*,
DOI 10.1007/978-1-4614-7098-4_1, © Springer Science+Business Media New York 2013

In our considerations, in general $\Omega = \mathbb{D}_R = \{z \in \mathbb{C}; |z| < R\}$ with $R > 1$, F is a segment included in $\mathbb{D}_R$, and the compact subsets considered will be the closed disks $\mathbb{D}_r = \{z \in \mathbb{C}; |z| \le r\}$ with $1 \le r < R$.

The second important result in complex analysis used is Cauchy's formula for disks.

Theorem 1.1.2 (Cauchy; see, e.g., Kohr–Mocanu [91], p. 28, Theorem 1.2.20). *Let $r > 0$ and $f : \overline{\mathbb{D}}_r \to \mathbb{C}$ be continuous in $\overline{\mathbb{D}}_r$ and analytic in $\mathbb{D}_r$. Then, for any $p \in \{0, 1, 2, \dots, \}$ and $|z| < r$, we have*

$$f^{(p)}(z) = \frac{p!}{2\pi i} \int_\Gamma \frac{f(u)}{(u-z)^{p+1}} du,$$

where $\Gamma = \{z \in \mathbb{C}; |z| = r\}$ and $i^2 = -1$.

An immediate consequence of Cauchy's formula is the following so-called Weierstrass's theorem.

Theorem 1.1.3 (Weierstrass; see, e.g., Kohr–Mocanu [91], p. 18, Theorem 1.1.6). *Let $G \subset \mathbb{C}$ be an open set. If the sequence $(f_n)_{n \in \mathbb{N}}$ of analytic functions in G converges uniformly in each compact of G to the analytic function f, then for any $p \in \mathbb{N}$, the sequence of pth derivatives $(f_n^{(p)})_{n \in \mathbb{N}}$ converges to $f^{(p)}$ uniformly on compacts in G.*

Indeed, by the above Cauchy's formula we can write

$$f_n^{(p)}(z) - f^{(p)}(z) = \frac{p!}{2\pi i} \int_\Gamma \frac{f_n(u) - f(u)}{(u-z)^{p+1}} du,$$

from which by passing to the modulus, the theorem immediately follows. In our considerations, we have $G = \mathbb{D}_R$ with $R > 1$, and the compact subsets in G are $\mathbb{D}_r$ with $1 \le r < R$.

Another well-known result is the following.

Theorem 1.1.4 (see, e.g., Graham–Kohr [74], Theorem 6.1.18). *If $f_n, f : \Omega \to \mathbb{C}$, $n \in \mathbb{N}$ are analytic functions in the domain Ω and f is univalent in Ω and $f_n \to f$ uniformly in the compact $K \subset \Omega$, then there exists an index $n_0(K)$, such that for all $n \ge n_0$, f_n is univalent in K.*

The classical Maximum Principle (or Maximum Modulus Theorem) will be frequently used in the proofs of error estimates.

Theorem 1.1.5 (see, e.g., Kohr–Mocanu [91], p. 2, Corollary 1.1.20). *Let $\Omega \subset \mathbb{C}$ be a bounded domain and $f : \overline{\Omega} \to \mathbb{C}$ be continuous in $\overline{\Omega}$ and analytic in Ω. Then, denoting by Γ the boundary of Ω, we have*

$$\max\{|f(z)|; z \in \overline{\Omega}\} = \max\{|f(z)|; z \in \Gamma\}.$$

For our considerations, Ω will be again an open disk with the center at origin.

Another useful result will be the well-known theorem on the zeroes of analytic functions, which in essence says that the zeroes of a nonidentical null analytic function necessarily are isolated points. More precisely, we can state the following.

Theorem 1.1.6 (see, e.g., Kohr–Mocanu [91], p. 20, Theorem 1.1.12). *Suppose that f is not identical null and analytic in the domain Ω. If a is a zero for f, then there is $r = r(a) > 0$ such that $\mathbb{D}(a,r) = \{z \in \mathbb{C}; |z - a| < r\} \subset \Omega$ and $f(z) \neq 0$, for all $z \in \mathbb{D}(a,r) \setminus \{a\}$.*

Also, we will use the so-called theorem on the identity of analytic functions.

Theorem 1.1.7 (see, e.g., Kohr–Mocanu [91], p. 21, Theorem 1.1.14). *Let $\Omega \subset \mathbb{C}$ be a domain. If $f, g : \Omega \to \mathbb{C}$ are analytic in Ω, then $f \equiv g$ on Ω is equivalent to the fact that the set $\{z \in \Omega; f(z) = g(z)\}$ has at least one accumulation point in Ω.*

Finally, we state a very basic result called Bernstein's inequality for complex polynomials in compact disks.

Theorem 1.1.8 (Bernstein [20], p. 45, relation (80) For general $r > 0$, see also, e.g., Lorentz [97], p. 40, Theorem 4, for $r = 1$). *Let $P(z) = \sum_{k=0}^{n} a_k z^k$ be with $a_k \in \mathbb{C}$, for all $k \in \{0, 1, 2, \ldots,\}$ and for $r > 0$, denote $\|P_n\|_r = \max\{|P_n(z)|; |z| \leq r\}$:*

(i) For all $|z| \leq 1$ we have $|P_n'(z)| \leq n\|P_n\|_1$.
(ii) If $r > 0$, then for all $|z| \leq r$, we have $|P_n'(z)| \leq \frac{n}{r}\|P_n\|_r$.

One observes that (ii) immediately follows from (i) because denoting $Q_n(z) = P_n(rz)$, $|z| \leq 1$, by (i) applied to $Q_n(z)$, it is immediate that $r|P_n'(rz)| \leq n\|P_n\|_r$, for all $|z| \leq 1$, which proves (ii).

Concerning the approximation of analytic functions by sequences of complex polynomials, as can be seen in the next sections of this chapter and also in the next chapters, in the main results, one refers to approximation in compact disks centered at origin (in particular in the compact unit disk). The advantage consists in the fact that in these kinds of disks, constructive methods can be indicated. But clearly it is important to obtain approximation results in more general domains of the complex plane. One of the standard methods is based on the so-called Faber polynomials introduced by Faber [34], which allows to extend all the constructive methods from the closed unit disk to more general domains. The method is less constructive because a generally unknown mapping function (generated from the Riemann's mapping theorem) enters into considerations. For all the details on this method, see, e.g., the book of Gaier [38], pp. 42–54. Also, for other important contributions to the field of constructive complex approximation, see the book of Dzjadyk [33].

Definition 1.1.9. (i) $\gamma : [a, b] \to \mathbb{C}$ is called a Jordan curve if it is closed (i.e., $\gamma(a) = \gamma(b)$) and simple (i.e., injective). The length of the curve γ is defined by

$$L(\gamma) = \sup\{\sum_{i=1}^{n} |\gamma(t_i) - \gamma(t_{i-1})|; n \in \mathbb{N}, a = t_0 < \ldots < t_n = b\}.$$

If $L < +\infty$, then γ is called rectifiable.

The interior of a Jordan curve is called Jordan domain and the corresponding curve is called boundary curve of that domain.

(ii) (Radon [124]) Suppose that $\gamma : [a, b] \to \mathbb{C}$ is a rectifiable Jordan curve. Because $L < +\infty$, it is known that γ has a tangent γ' almost everywhere. Then γ is called of bounded rotation if γ' can be extended to a function of bounded variation on the whole curve.

Remark. Simple examples of Jordan curve of bounded rotation can be made up of finitely many convex arcs (where corners are permitted).

Now, if G is a Jordan domain, then (by the Riemann's mapping theorem) let us denote by Ψ the conformal mapping of $\mathbb{C} \setminus \overline{\mathbb{D}}_1$ onto $\mathbb{C} \setminus \overline{G}$, normalized at ∞, that is, $0 < \lim_{w \to \infty} \frac{\Psi(w)}{w} < \infty$. Also, denote by Φ the inverse function of Ψ. Obviously that Ψ and Φ depend on $\overline{G}$, but for the simplicity of notation we will not write them as $\Psi_{\overline{G}}$ and $\Phi_{\overline{G}}$, considering in our presentation that $\overline{G}$ is arbitrary but fixed.

For a Jordan domain G, denote by $A(\overline{G})$ the class of all functions analytic in G and continuous in $\overline{G}$. In what follows we sketch a method by which any $f \in A(\overline{G})$ can be approximated by polynomials. For our considerations, it is good enough to suppose that the boundary curve of G is rectifiable and of bounded rotation.

Firstly, let us consider the Laurent expansion of $[\Phi(z)]^n$, $n \in \mathbb{N} \cup \{0\}$, valid for large z:

$$[\Phi(z)]^n = a_0^{(n)} + \ldots + a_n^{(n)} z^n + \sum_{k=1}^{\infty} a_{-k}^{(n)}/z^k.$$

Definition 1.1.10 (Faber [34]).

(i) The polynomial $F_n(z) = a_0^{(n)} + \ldots + a_n^{(n)} z^n$, $n \in \mathbb{N} \cup \{0\}$ is called the Faber polynomial of degree n attached to the domain G. (Note that for $z \in \mathbb{D}_R$ and $R > 1$, we can write

$$F_n(z) = \frac{1}{2\pi i} \int_{|u|=R} \frac{[\Phi(u)]^n}{u - z} du.)$$

(ii) Let $f \in A(\overline{G})$. The complex numbers

$$a_n(f) = \frac{1}{2\pi i} \int_{|u|=1} \frac{f[\Psi(u)]}{u^{n+1}} du = \frac{1}{2\pi i} \int_{-\pi}^{\pi} f[\Psi(e^{it})] e^{-int} dt, n \in \mathbb{N} \cup \{0\},$$

are called the Faber coefficients of f, and $\sum_{n=0}^{\infty} a_n(f)F_n(z)$ is called the Faber series attached to f on G. The Faber series represent a natural generalization of the Taylor series in the case when the unit disk is replaced by an arbitrary simply connected domain bounded by a "nice" curve.

1.2 Voronovskaja-Type Theorems for Derivatives of Bernstein Polynomials

In this section we obtain the differentiated generalized Voronovskaja's theorem in complex setting with upper and exact quantitative estimates. The results extend that obtained in the real case on $[0, 1]$ in Gonska–Raşa [72] and generalize those obtained in the complex case in Gal [46]; see also the book of Gal [49], pp. 36–42, Theorem 1.3.2 and Corollary 1.3.4.

For $f : [0, 1] \to \mathbb{R}$, let us define the sequence of real Bernstein polynomials attached to f by $B_n(f)(x) = \sum_{k=0}^{n} \binom{n}{k} x^k (1-x)^{n-k} f(k/n)$, $n \in \mathbb{N}, x \in [0, 1]$.

Recently, the following quantitative estimate for the differentiated Voronovskaja's formula for the real Bernstein polynomials has been obtained.

Theorem 1.2.1 (Gonska–Raşa [72]). *If $f \in C^{k+2}[0, 1]$ for some $k \geq 0$, then for all $x \in [0, 1]$ and $n \in \mathbb{N}$, we have*

$$\left| n[(B_n(f))^{(k)}(x) - f^{(k)}(x)] - \frac{1}{2} \cdot \frac{d^k}{dx^k} \{x(1-x)f''(x)\} \right| \leq$$

$$\frac{C_k}{n} \cdot \max_{k \leq j \leq k+2} \{|f^{(j)}(x)|\} + M_k \cdot \tilde{\omega}_1 \left(f^{(k+2)}; \frac{1}{\sqrt{n}} \right),$$

where $C_k, M_k > 0$ and $\tilde{\omega}_1(f; \delta)$ is the least concave majorant of the usual modulus of continuity $\omega_1(f; \delta)$ satisfying the inequalities $\omega_1(f; \delta) \leq \tilde{\omega}_1(f; \delta) \leq 2\omega_1(f; \delta)$ for all $\delta \geq 0$.

Now, if we denote

$$T_{n,j}(x) := \sum_{k=0}^{n} (k - nx)^j \binom{n}{k} x^k (1-x)^{n-k},$$

then it is known that the Voronovskaja's theorem in Voronovskaja [142] was generalized by Bernstein as follows.

Theorem 1.2.2 (Bernstein [19]; see, e.g., Lorentz [96], p. 22–23). *Let $f : [0, 1] \to \mathbb{C}$ be a bounded function such that the derivative $f^{(2p)}(x)$ exists at x. Then*

$$B_n(f)(x) = f(x) + \sum_{j=0}^{2p} \frac{f^{(j)}(x)}{j!} n^{-j} T_{n,j}(x) + \frac{\varepsilon_n}{n^p}$$

where $\varepsilon_n \to 0$ as $n \to \infty$.

This theorem was extended to analytic functions in a disk $\mathbb{D}_R = \{z \in \mathbb{C}; |z| < R\}$, by the following result.

Theorem 1.2.3 (Gal [46]). *Let $R > 1$ and $f : \mathbb{D}_R \to \mathbb{C}$ be an analytic function in $\mathbb{D}_R$, that is, $f(z) = \sum_{k=0}^{\infty} c_k z^k$ for all $z \in \mathbb{D}_R$. Then for any $1 \leq r < R$ and any natural number p, there exists a constant $C_p > 0$ such that we have*

$$\left\| B_n(f) - f - \sum_{j=1}^{2p} \frac{f^{(j)}}{j!} n^{-j} T_{n,j} \right\|_r \leq \frac{C_{p,r}(f)}{n^{p+1}},$$

for all $n \in \mathbb{N}$, where

$$C_{p,r}(f) = C_p \cdot \sum_{j=2p+1}^{\infty} |c_j| \frac{j!(j-2p)}{(j-2p-1)!} r^j < \infty.$$

Here $\|f\|_r = \sup_{|z| \leq r} |f(z)|$.

Moreover, the following exact estimate has been obtained.

Theorem 1.2.4 (Gal [46]). *Let $R > 1$ and let $f : \mathbb{D}_R \to \mathbb{C}$ be an analytic function in $\mathbb{D}_R$, that is, $f(z) = \sum_{k=0}^{\infty} c_k z^k$ for all $z \in \mathbb{D}_R$. If f is not a polynomial of degree $\leq 2p$, then for any $1 \leq r < R$ and any natural number p, we have*

$$\left\| B_n(f) - f - \sum_{j=1}^{2p} \frac{f^{(j)}}{j!} n^{-j} T_{n,j} \right\|_r \sim \frac{1}{n^{p+1}}, \quad n \in \mathbb{N},$$

where the constants in the equivalence depend only on f, r, and p and are independent of n.

The goal of this section is to present upper and exact quantitative estimates for

$$\left\| \left[B_n(f) - f - \sum_{j=1}^{2p} \frac{f^{(j)}}{j!} n^{-j} T_{n,j} \right]^{(k)} \right\|_r,$$

generalizing thus the Theorems 1.2.1, 1.2.3, and 1.2.4.

The following upper estimate holds.

Theorem 1.2.5 (Gal [53]). *Let $1 \leq r < r_1 < R$ and $f : \mathbb{D}_R \to \mathbb{C}$ be an analytic function in $\mathbb{D}_R$, that is, $f(z) = \sum_{k=0}^{\infty} c_k z^k$, for all $z \in \mathbb{D}_R$. Then for $p, k \in \mathbb{N}$, there exists a constant $C_{p,r_1}(f) > 0$ such that the upper estimate*

$$\left\| \left[B_n(f) - f - \sum_{j=1}^{2p} \frac{f^{(j)}}{j!} n^{-j} T_{n,j} \right]^{(k)} \right\|_r \leq C_{p,r_1}(f) \cdot \frac{k! r_1}{n^{p+1}(r_1 - r)^{k+1}},$$

holds for all $n \in \mathbb{N}$, where

$$C_{p,r_1}(f) = C_p \cdot \sum_{j=2p+1}^{\infty} |c_j| \frac{j!(j-2p)}{(j-2p-1)!} r_1^j < \infty.$$

Proof. Denote $F(z) = B_n(f)(z) - f(z) - \sum_{j=1}^{2p} \frac{f^{(j)}(z)}{j!} n^{-j} T_{n,j}(z)$. By Cauchy's theorem it follows

$$F^{(k)}(z) = \frac{k!}{2\pi i} \int_{\Gamma_1} \frac{F(v)}{(v-z)^{k+1}} dv,$$

where Γ_1 is the circle of center 0 and radius r_1.

Passing to supremum with $|z| \leq r$, we get

$$\|F^{(k)}\|_r \leq \frac{k! 2\pi r_1}{2\pi} \cdot \frac{\|F\|_{r_1}}{(r_1-r)^{k+1}},$$

which, together with the estimate for $\|F\|_{r_1}$ in Theorem 1.2.3, immediately implies the desired estimate. $\qquad\square$

In the next results, one refers to the exact estimates.

Theorem 1.2.6 (Gal [53]). *Let $R > 1$, $p, k \in \mathbb{N}$ and $f : \mathbb{D}_R \to \mathbb{C}$ be an analytic function in $\mathbb{D}_R$, that is, $f(z) = \sum_{k=0}^{\infty} c_k z^k$, for all $z \in \mathbb{D}_R$.*
Let $k < 2p + 1$. If f is not a polynomial of degree $\leq 2p$, then for any $1 \leq r < R$, we have

$$\left\| \left[B_n(f) - f - \sum_{j=1}^{2p} \frac{f^{(j)}}{j!} n^{-j} T_{n,j} \right]^{(k)} \right\|_r \sim \frac{1}{n^{p+1}}, \quad n \in \mathbb{N},$$

where the constants in the equivalence depend only on f, r, p, k but are independent of n.

Proof. By Theorem 1.2.5, it remains to prove the lower estimate.

From Gal [46], p. 263, we can write

$$B_n(f)(z) - f(z) - \sum_{j=1}^{2p} \frac{f^{(j)}(z)}{j!} n^{-j} T_{n,j}(z)$$

$$= \frac{1}{n^{p+1}} \left\{ \frac{a_p}{(2p+1)!} (1-2z)[z(1-z)]^p f^{(2p+1)}(z) + \frac{[z(1-z)]^{p+1}}{2^{p+1}(p+1)!} f^{(2p+2)}(z) \right.$$

$$+ \frac{1}{n} F(z) f^{(2p+1)}(z) + \frac{1}{n} G(z) f^{(2p+2)}(z) + \frac{1}{n} \left[n^{p+2} \sum_{j=2p+3}^{\infty} c_j E_{j,n,p+1}(z) \right] \right\},$$

where $a_p > 0$,

$$E_{j,n,p}(z) = B_n(e_j)(z) - e_j(z) - \sum_{q=1}^{2p} \frac{1}{q!n^q} T_{n,q}(z)(z^j)^{(q)}, e_j(z) = z^j,$$

and the polynomials $F(z) := P_1(z) + \frac{1}{n}P_2(z) + \ldots + \frac{1}{n^{p-1}}P_p(z)$, $G(z) := Q_1(z) + \frac{1}{n}Q_2(z) + \ldots + \frac{1}{n^p}Q_{p+1}(z)$ are bounded in any closed disk $|z| \le r$ by constants depending on r and p but independent of n.

Taking in the above identity the derivative of order k and by the following inequalities

$$\|h + g\|_r \ge |\,\|h\|_r - \|g\|_r\,| \ge \|h\|_r - \|g\|_r,$$

we get

$$\left\|\left[B_n(f) - f - \sum_{j=1}^{2p} \frac{f^{(j)}}{j!}n^{-j}T_{n,j}\right]^{(k)}\right\|_r \ge$$

$$\frac{1}{n^{p+1}}\left\{\left\|\left[\frac{a_p}{(2p+1)!}(1-2e_1)[e_1(1-e_1)]^p f^{(2p+1)}\right.\right.\right.$$

$$\left.\left.+\frac{[e_1(1-e_1)]^{p+1}}{2^{p+1}(p+1)!}f^{(2p+2)}\right]^{(k)}\right\|_r$$

$$-\frac{1}{n}\left[\left\|n^{p+2}\sum_{j=2p+3}^{\infty} c_j E_{j,n,p+1}^{(k)} + \left[f^{2p+1}F + f^{(2p+2)}G\right]^{(k)}\right\|_r\right]\right\}$$

$$:= \frac{1}{n^{p+1}}\left\{\|U\|_r - \frac{1}{n}[\|V\|_r]\right\} \ge \frac{1}{n^{p+1}} \cdot \frac{1}{2}\|U\|_r,$$

for all $n \ge n_0$ (n_0 depends on f, p, k and r), under the conditions that $\|U\|_r > 0$ and $\|V\|_r$ is upper bounded by a constant depending only on f, p, and r.

But this is exactly the case. Indeed, first we observe that by the formulas (2.3) and (2.4) in Gal [46], we can write

$$\left[B_n(f)(z) - f(z) - \sum_{q=1}^{2p+2} \frac{f^{(q)}(z)}{q!}n^{-q}T_{n,q}(z)\right]^{(k)} = \sum_{j=2p+3}^{\infty} c_j E_{j,n,p+1}^{(k)}(z),$$

which, from Cauchy's formula and from the estimate in Theorem 1.1 in Gal [46] (written for $p+1$), immediately implies that $n^{p+2}\|\sum_{j=2p+3}^{\infty} c_j E_{j,n,p+1}^{(k)}\|_r$ is upper bounded by a positive constant depending only on f, p, k and r.

Combined with the above considerations on F and G, it immediately follows that $\|V\|_r$ is upper bounded by a constant depending only on f, p, k and r.

What remains to prove is that $\|U\|_r > 0$. Indeed, supposing the contrary, we get that f necessarily satisfies the differential equation

$$\left[\frac{a_p}{(2p+1)!}(1-2z)[z(1-z)]^p f^{(2p+1)}(z) + \frac{[z(1-z)]^{p+1}}{2^{p+1}(p+1)!} f^{(2p+2)}(z)\right]^{(k)} = 0,$$

for all $|z| \le r$.

By the substitution $f^{(2p+1)}(z) := y(z)$, we obtain that $y(z)$ necessarily is analytic in $\mathbb{D}_R$ (since f is supposed analytic there) and is a solution of the differential equation

$$\frac{a_p}{(2p+1)!}(1-2z)[z(1-z)]^p y(z) + \frac{[z(1-z)]^{p+1}}{2^{p+1}(p+1)!} y'(z) = P_{k-1}(z), |z| \le r,$$

where $P_{k-1}(z)$ is a polynomial of degree $\le k-1 < 2p$.

Dividing by $[z(1-z)]^p$ in the above differential equation, we get that the analytic function $y(z)$ satisfies the differential equation (recall that $a_p > 0$)

$$\frac{a_p}{(2p+1)!}(1-2z)y(z) + \frac{z(1-z)}{2^{p+1}(p+1)!} y'(z) = \frac{P_{k-1}(z)}{[z(1-z)]^p}, |z| \le r,$$

which is impossible if $P_{k-1}(z)$ is not identical zero. Indeed, in this case the left-hand side is analytic in $\mathbb{D}_R$, while the right-hand side has poles at $z = 0$ and $z = 1$. Therefore, this means that we necessarily have $P_{k-1}(z) = 0$ for all $|z| \le r$ and we get the differential equation

$$\frac{a_p}{(2p+1)!}(1-2z)[z(1-z)]^p y(z) + \frac{[z(1-z)]^{p+1}}{2^{p+1}(p+1)!} y'(z) = 0, |z| \le r.$$

Now, reasoning exactly as in the proof of Corollary 1.2, at page 264 in Gal [46], it necessarily follows that f is a polynomial of degree $\le 2p$, which is a contradiction with the hypothesis.

In conclusion, we necessarily have $\|U\|_r > 0$.

Now, for $n \in \{1, \ldots, n_0 - 1\}$, we obviously get

$$\left\|\left[B_n(f) - f - \sum_{j=1}^{2p} \frac{f^{(j)}}{j!} n^{-j} T_{n,j}\right]^{(k)}\right\|_r \ge \frac{M_{r,n}(f)}{n^{p+1}},$$

with $M_{r,n}(f) = n^{p+1} \cdot \left\|\left[B_n(f) - f - \sum_{j=1}^{2p} \frac{f^{(j)}}{j!} n^{-j} T_{n,j}\right]^{(k)}\right\|_r > 0$, which finally implies

$$\left\|\left[B_n(f) - f - \sum_{j=1}^{2p} \frac{f^{(j)}}{j!} n^{-j} T_{n,j}\right]^{(k)}\right\|_r \geq \frac{C_{p,r}(f)}{n^{p+1}}, \text{ for all } n \in \mathbb{N},$$

where $C_{p,r}(f) = \min\{M_{r,1}(f), \ldots, M_{r,n_0-1}(f), \frac{1}{2}\|U\|_r\}$. $\qquad\qquad\square$

Remark. In the paper of Gal [53], Theorem 1.1.6 was stated without restriction for $p, k \in \mathbb{N}$ and under the hypothesis that f is not a polynomial of degree $\leq \max\{k-1, 2p\}$. Unfortunately, while this statement is valid for $k < 2p+1$ by Theorem 1.2.6, it seems to be not valid, in general, for $k \geq 2p+1$, as the following counterexample shows. For $p = 1$ and $k = 3$ (that is for $k = 2p+1$), choose $f(z) = z^3$. The question is to verify that

$$\left\|\left[B_n(f) - f - \frac{e_1(1-e_1)f''}{n}\right]'''\right\|_r \sim \frac{1}{n^2},$$

does not hold. Indeed, this is the case because by $B_n(e_3)(z) = z^3 + \frac{3z^2(1-z)}{n} + \frac{z(1-z)(1-2z)}{n^2}$, a simple calculation implies

$$\left\|\left[B_n(e_3) - f - \frac{e_1(1-e_1)e_3''}{n}\right]'''\right\|_r \sim \frac{1}{n} + \frac{1}{n^2} \sim \frac{1}{n}.$$

However, analyzing the case $k \geq 2p+1$ in the proof of Theorem 2.2 in Gal [53], it can be recovered under some simple additional hypothesis, that is, we can deduce that for some large subclasses of analytic functions, the lower estimate of order $\frac{1}{n^{p+1}}$ still holds for $k \geq 2p+1$ too. In this sense, as a sample we present the following.

Theorem 1.2.7 (Gal [63]). *Let $R > 1$, $p, k \in \mathbb{N}$ and $f : \mathbb{D}_R \to \mathbb{C}$ be an analytic function in $\mathbb{D}_R$, that is, $f(z) = \sum_{k=0}^{\infty} c_k z^k$, for all $z \in \mathbb{D}_R$.*

(i) Let $k = 2p+1$. If f is not a polynomial of degree $\leq k-1$ and $\{f^{(2p+1)}(0) = 0$ or $f^{(2p+1)}(1) = 1\}$, then for any $1 \leq r < R$, we have

$$\left\|\left[B_n(f) - f - \sum_{j=1}^{2p} \frac{f^{(j)}}{j!} n^{-j} T_{n,j}\right]^{(k)}\right\|_r \sim \frac{1}{n^{p+1}}, \quad n \in \mathbb{N},$$

where the constants in the equivalence depend only on f, r, p, k but are independent of n.

(ii) Let $k = 2p+2$ or $k = 2p+3$. If f is not a polynomial of degree $\leq k-1$ and $\{f^{(2p+1)}(0) = f^{(2p+1)}(1) = f^{(2p+2)}(1/2) = 0\}$, then again the equivalence from the above point (i) holds.

(iii) More general, let $k = 2p+s$ with $s \geq 4$. If f is not a polynomial of degree $\leq k-1$ and

$$f^{(2p+1)}(0) = f^{(2p+2)}(0) = \ldots = f^{(2p+s-1)}(0) = f^{(2p+2)}(1/2) = f^{(2p+1)}(1) = 0,$$

or

$$f^{(2p+1)}(1) = f^{(2p+2)}(1) = \ldots = f^{(2p+s-1)}(1) = f^{(2p+2)}(1/2) = f^{(2p+1)}(0) = 0,$$

then again the equivalence from the above point (i) holds.

Proof. By Theorem 1.2.5, in the above points (i) and (ii), what remains to be proved are the lower estimates.

Keeping the notations in the proof of Theorem 1.2.6, what remains to be proved is that $\|U\|_r > 0$. Indeed, supposing the contrary, we get that f necessarily satisfies the differential equation

$$\left[\frac{a_p}{(2p+1)!}(1-2z)[z(1-z)]^p f^{(2p+1)}(z) + \frac{[z(1-z)]^{p+1}}{2^{p+1}(p+1)!} f^{(2p+2)}(z) \right]^{(k)} = 0,$$

for all $|z| \leq r$. By the substitution $f^{(2p+1)}(z) := y(z)$, we obtain that $y(z)$ necessarily is analytic in $\mathbb{D}_R$ (since f is supposed analytic there) and is a solution of the differential equation

$$\frac{a_p}{(2p+1)!}(1-2z)[z(1-z)]^p y(z) + \frac{[z(1-z)]^{p+1}}{2^{p+1}(p+1)!} y'(z) = P_{k-1}(z), |z| \leq r,$$

where $P_{k-1}(z)$ is a polynomial of degree $\leq k-1$.

Since $k \geq 2p+1$, we necessarily obtain that $P_{k-1}(z) = [z(1-z)]^p Q_l(z)$, where $l = k-1-2p \geq 0$ and $Q_l(z)$ is a polynomial of degree $\leq l$ (contrariwise, we again would get that $P_{k-1}(z)$ must be identical zero and f would be a polynomial of degree $\leq 2p \leq k-1$, which would be a contradiction). Dividing by $[z(1-z)]^p$, we obtain that $y(z)$ is an analytic function in $\mathbb{D}_R$, satisfying the differential equation (here recall that $a_p > 0$)

$$\frac{a_p}{(2p+1)!}(1-2z)y(z) + \frac{z(1-z)}{2^{p+1}(p+1)!} y'(z) = Q_l(z), |z| \leq r, z \neq 0, z \neq 1.$$
$$(1.2.1)$$

(i) Let $k = 2p+1$, that is, above we have $l = 0$ and $Q_l(z)$ is a constant. By hypothesis, in (1.2.1) we have $y(0) = 0$ or $y(1) = 0$, which implies $Q_l(0) = 0$ or $Q_l(1) = 0$, that is, (1.2.1) necessarily becomes

$$\frac{a_p}{(2p+1)!}(1-2z)y(z) + \frac{z(1-z)}{2^{p+1}(p+1)!} y'(z) = 0, |z| \leq r.$$

Reasoning exactly as in the proof of Corollary 1.2, at page 164 in Gal [46], that is, writing $y(z)$ in the form $y(z) = \sum_{k=0}^{\infty} b_k z^k$, by comparison of coefficients, we easily obtain that $b_k = 0$, for all $k = 0, 1, \ldots$, which implies that $y(z)$ is identical zero in $\overline{\mathbb{D}}_r$. But from the identity theorem of analytic functions, it necessarily follows that $y(z) = 0$ for all $|z| < R$. It follows that $f(z)$ necessarily is a polynomial of degree $\leq 2p = k - 1$, a contradiction with the hypothesis.

(ii) Let $k = 2p + 2$ or $k = 2p + 3$, that is, $l = 1$ or $l = 2$ in the differential Equation (1.2.1). By the hypothesis it follows $Q_l(0) = Q_l(1) = Q_l(1/2) = 0$, where $Q_l(z)$ is a polynomial of degree ≤ 2, which necessarily implies that Q_l is identically equal to zero. In continuation, reasoning exactly as in the proof of the above point (i), it necessarily follows that $y(z) = 0$ for all $|z| < R$, that is, $f(z)$ necessarily is a polynomial of degree $\leq 2p < k - 1$, contradicting the hypothesis.

(iii) Let $k = 2p + s$ with $s \geq 4$. It follows that in the differential equation (1.2.1), we have $l = s - 1$. Differentiating successively (1.2.1) until $s - 3$ (including $s - 3$), it is easy to check that we get the equalities of the form

$$c_1^{(1)} y(z) + c_2^{(1)}(1 - 2z)y'(z) + c_3^{(1)} z(1 - z)y''(z) = Q_l'(z),$$

$$c_1^{(2)} y'(z) + c_2^{(2)}(1 - 2z)y''(z) + c_3^{(2)} z(1 - z)y'''(z) = Q_l''(z),$$

and so on, until

$$c_1^{(s-3)} y^{(s-4)}(z) + c_2^{(s-3)}(1 - 2z)y^{(s-3)}(z) + c_3^{(s-3)} z(1 - z)y^{(s-2)}(z) = Q_l^{(s-3)}(z),$$

where $c_k^{(j)}$ are real constants.

Taking now into account the hypothesis, we immediately obtain

$$Q_l(0) = Q_l'(0) = \ldots = Q_l^{(s-3)}(0) = 0, \; Q_l(1/2) = 0, \; Q_l(1) = 0,$$

or

$$Q_l(1) = Q_l'(1) = \ldots = Q_l^{(s-3)}(1) = 0, \; Q_l(1/2) = 0, \; Q_l(0) = 0,$$

respectively, which immediately implies that Q_l is identically equal to zero. In continuation, reasoning exactly as in the proof of the above point (i), it necessarily follows that $y(z) = 0$ for all $|z| < R$, that is, $f(z)$ necessarily is a polynomial of degree $\leq 2p < k - 1$, contradicting the hypothesis.

So in all the three cases (i), (ii), and (iii) we necessarily have $\|U\|_r > 0$.

For $n \in \{1, \ldots, n_0 - 1\}$, in all the three cases (i), (ii), and (iii) we obviously get

$$\left\| \left[B_n(f) - f - \sum_{j=1}^{2p} \frac{f^{(j)}}{j!} n^{-j} T_{n,j} \right]^{(k)} \right\|_r \geq \frac{M_{r,n}(f)}{n^{p+1}},$$

with $M_{r,n}(f) = n^{p+1} \cdot \left\| \left[B_n(f) - f - \sum_{j=1}^{2p} \frac{f^{(j)}}{j!} n^{-j} T_{n,j} \right]^{(k)} \right\|_r > 0$, which finally implies

$$\left\| \left[B_n(f) - f - \sum_{j=1}^{2p} \frac{f^{(j)}}{j!} n^{-j} T_{n,j} \right]^{(k)} \right\|_r \geq \frac{C_{p,r}(f)}{n^{p+1}}, \text{ for all } n \in \mathbb{N},$$

where $C_{p,r}(f) = \min\{M_{r,1}(f), \ldots, M_{r,n_0-1}(f), \frac{1}{2}\|U\|_r\}$. □

Remark. Simple functions f satisfying the hypothesis of Theorem 1.2.7 are of the form $f(z) = z^{2p+s}(1-z)^{2p+2}(z-1/2)^{2p+3}g(z)$ or $f(z) = (1-z)^{2p+s}z^{2p+2}(z-1/2)^{2p+3}g(z)$, where $s \geq 4$ and g is an arbitrary not zero analytic function in $\mathbb{D}_R$. We see that $f(z) = z^3$ does not satisfy any hypothesis in Theorem 1.2.7.

1.3 Schurer–Faber Polynomials

In this section, approximation properties in compact sets of the complex plane by the so-called Schurer–Faber polynomials are presented.

Let us recall that concerning the convergence of Bernstein polynomials in the complex plane, Bernstein proved (see, e.g., Lorentz [96], p. 88) that if $f : G \to \mathbb{C}$ is analytic in the open set $G \subset \mathbb{C}$, with $\overline{\mathbb{D}_1} \subset G$ (where $\mathbb{D}_1 = \{z \in \mathbb{C} : |z| < 1\}$), then the complex Bernstein polynomials $B_n(f)(z) = \sum_{k=0}^{n} \binom{n}{k} z^k (1-z)^{n-k} f(k/n)$ uniformly converge to f in $\overline{\mathbb{D}_1}$.

Exact estimates of order $O(1/n)$ of this uniform convergence and, in addition, of the simultaneous approximation were found in Gal [43] and Gal [39]. In Gal [41] a Voronovskaja-type result with quantitative estimate for complex Bernstein polynomials in compact disks and shape-preserving properties in the unit disk were obtained. Extensions to approximation in compact sets of $\mathbb{C}$ by the so-called Bernstein–Faber polynomials were proved in Gal [49], pp. 19–25.

Also, in Gal [44, 45, 47] similar results for complex Bernstein–Stancu and Kantorovich–Stancu polynomials were obtained.

In this section, we will deal with the so-called Schurer–Faber polynomials attached to compact sets in $\mathbb{C}$. In the particular case when $p = 0$, we will recapture the results for the complex Bernstein–Faber polynomials obtained in the book of Gal [49], pp. 19–25.

Here $G \subset \mathbb{C}$ will be considered a compact set such that $\tilde{\mathbb{C}} \setminus G$ is connected. In this case, according to the Riemann mapping theorem, a unique conformal mapping Ψ of $\tilde{\mathbb{C}} \setminus \overline{\mathbb{D}_1}$ onto $\tilde{\mathbb{C}} \setminus G$ exists so that $\Psi(\infty) = \infty$ and $\Psi'(\infty) > 0$.

By using the Faber polynomials $F_p(z)$ attached to G (see Definition 1.1.10) and starting from the the complex Bernstein–Schurer polynomials (introduced and studied in the case of real variable in Schurer [130]) defined for any fixed $p \in \{0, 1, 2, \ldots\}$ by

$$S_{n,p}(f)(z) = \sum_{k=0}^{n+p} \binom{n+p}{k} z^k (1-z)^{n+p-k} f(k/n), z \in \mathbb{C},$$

for $f \in A(\overline{G})$, we can introduce the Schurer–Faber polynomials given by the formula

$$S_{n,q}(f; \overline{G})(z) = \sum_{p=0}^{n+q} \binom{n+q}{p} \Delta_{1/n}^p F(0) \cdot F_p(z), \ z \in G, \ n \in \mathbb{N},$$

where

$$\Delta_h^p F(0) = \sum_{k=0}^{p} (-1)^{p-k} \binom{p}{k} F(kh), \ F(w) = \frac{1}{2\pi i} \int_{|u|=1} \frac{f(\Psi(u))}{u - w} du, \ w \in \mathbb{D}_1.$$

Here, since $F(1)$ is involved in $\Delta_{1/n}^n F(0)$ and therefore in the definition of $S_{n,q}(f; G)(z)$ too, in addition we will suppose that F can be extended by continuity on the boundary $\partial \mathbb{D}_1$.

Remark. 1) For $G = \overline{\mathbb{D}}_1$, it is easy to see that in the above Schurer–Faber polynomials, one reduces to the classical complex Bernstein–Schurer polynomials (studied in Anastassiou and Gal [9]) given by

$$S_{n,q}(f)(z) = \sum_{p=0}^{n+q} \binom{n+q}{p} \Delta_{1/n}^p f(0) z^p = \sum_{p=0}^{n+q} \binom{n+q}{p} z^p (1-z)^{n+q-p} f(p/n).$$

2) It is known that, for example, $\int_0^1 \frac{\omega_p(f \circ \Psi; u)_{\partial \mathbb{D}_1}}{u} du < \infty$ is a sufficient condition for the continuity on $\partial \mathbb{D}_1$ of F in the above definition of the Schurer–Faber polynomials (see, e.g., Gaier [38], p. 52, Theorem 6). Here $p \in \mathbb{N}$ is arbitrary fixed.

In the first main result, one refers to approximation on compact sets without any restriction on their boundaries and can be stated as follows.

Theorem 1.3.1. *Let $q \in \mathbb{N} \cup \{0\}$ be fixed and G be a continuum (i.e., a connected compact subset of $\mathbb{C}$) and suppose that f is analytic in G such that there exists $R > q + 1$ with f being analytic in G_R. Here recall that G_R denotes the interior of the closed level curve Γ_R given by $\Gamma_R = \{z; |\Phi(z)| = R\} = \{\Psi(w); |w| = R\}$ (and that $G \subset \overline{G}_r$ for all $1 < r < R$). Also, we suppose that F given in the definition of Schurer–Faber polynomials can be extended by continuity on $\partial \mathbb{D}_1$.*

For any R and r with $1 \leq r(q + 1) < R$ the following estimate

$$|S_{n,q}(f; \overline{G})(z) - f(z)| \leq \frac{C}{n}, \ z \in \overline{G}_r, \ n \in \mathbb{N},$$

holds, where $C > 0$ depends on f, q, r, and G_r but it is independent of n.

Proof. First we note that since G is a continuum, then it follows that $\tilde{\mathbb{C}} \setminus G$ is simply connected. By the proof of Theorem 2, p. 52–53 in Suetin [136] (by choosing there $K = \overline{G}_\beta$), for any fixed $1 < \beta(q+1) < R$, we have $f(z) = \sum_{k=0}^{\infty} a_k(f)F_k(z)$ uniformly in $\overline{G}_\beta$, where $a_k(f)$ are the Faber coefficients and are given by $a_k(f) = \frac{1}{2\pi i} \int_{|u|=\beta} \frac{f[\Psi(u)]}{u^{k+1}} du$. Note here that $G \subset \overline{G}_\beta$.

First we will prove that

$$S_{n,q}(f; \overline{G})(z) = \sum_{k=0}^{\infty} a_k(f)S_{n,q}(F_k; \overline{G})(z),$$

for all $z \in G$. (Note here that by hypothesis we have $\overline{G} = G$.) For this purpose, denote $f_m(z) = \sum_{k=0}^{m} a_k(f)F_k(z)$, $m \in \mathbb{N}$.

Since by the linearity of $S_{n,q}$, we easily get

$$S_{n,q}(f_m; \overline{G})(z) = \sum_{k=0}^{m} a_k(f)S_{n,q}(F_k; \overline{G})(z), \text{ for all } z \in G,$$

it suffices to prove that $\lim_{m\to\infty} S_{n,q}(f_m; \overline{G})(z) = S_{n,q}(f; \overline{G})(z)$, for all $z \in G$ and $n \in \mathbb{N}$.

First we have

$$S_{n,q}(f_m; \overline{G})(z) = \sum_{p=0}^{n+q} \binom{n+q}{p} \Delta_{1/n}^p G_m(0)F_p(z),$$

where $G_m(w) = \frac{1}{2\pi i} \int_{|u|=1} \frac{f_m(\Psi(u))}{u-w} du$ and $F(w) = \frac{1}{2\pi i} \int_{|u|=1} \frac{f(\Psi(u))}{u-w} du$.

Note here that since by Gaier [38], p. 48, the first relation before (6.17), we have

$$\mathcal{F}_k(w) = \frac{1}{2\pi i} \int_{|u|=1} \frac{F_k(\Psi(u))}{u-w} du = w^k, \text{ for all } |w| < 1,$$

evidently that $\mathcal{F}_k(w)$ can be extended by continuity on $\partial \mathbb{D}_1$. This also immediately implies that $G_m(w) = \frac{1}{2\pi i} \int_{|u|=1} \frac{f_m(\Psi(u))}{u-w} du$ can be extended by continuity on $\partial \mathbb{D}_1$, which means that $S_{n,q}(F_k; G)(z)$ and $S_{n,q}(f_m; G)(z)$ are well defined.

Now, taking into account Cauchy's theorem we also can write

$$G_m(w) = \frac{1}{2\pi i} \int_{|u|=\beta} \frac{f_m(\Psi(u))}{u-w} du \text{ and } F(w) = \frac{1}{2\pi i} \int_{|u|=\beta} \frac{f(\Psi(u))}{u-w} du.$$

For all $n, m \in \mathbb{N}$ and $z \in G$ it follows

$$|S_{n,q}(f_m; \overline{G})(z) - S_{n,q}(f; \overline{G})(z)|$$
$$\leq \sum_{p=0}^{n+q} \binom{n+q}{p} |\Delta_{1/n}^p(G_m - F)(0)| \cdot |F_p(z)|$$

$$\leq \sum_{p=0}^{n+q} \binom{n+q}{p} \sum_{j=0}^{p} \binom{p}{j} |(G_m - F)((p-j)/n)| \cdot |F_p(z)|$$

$$\leq \sum_{p=0}^{n+q} \binom{n+q}{p} \sum_{j=0}^{p} \binom{p}{j} C_{j,p,\beta} \|f_m - f\|_{\overline{G}_\beta} \cdot |F_p(z)|$$

$$\leq M_{n,q,\beta,G_\beta} \|f_m - f\|_{\overline{G}_\beta},$$

which by $\lim_{m \to \infty} \|f_m - f\|_{\overline{G}_\beta} = 0$ (see, e.g., the proof of Theorem 2, p. 52 in Suetin [136]) implies the desired conclusion. Here $\|f_m - f\|_{\overline{G}_\beta}$ denotes the uniform norm of $f_m - f$ on $\overline{G}_\beta$.

Consequently we obtain

$$|S_{n,q}(f; \overline{G})(z) - f(z)| \leq \sum_{k=0}^{\infty} |a_k(f)| \cdot |S_{n,q}(F_k; \overline{G})(z) - F_k(z)|$$

$$= \sum_{k=0}^{n+q} |a_k(f)| \cdot |S_{n,q}(F_k; \overline{G})(z) - F_k(z)|$$

$$+ \sum_{k=n+q+1}^{\infty} |a_k(f)| \cdot |S_{n,q}(F_k; \overline{G})(z) - F_k(z)|.$$

Therefore it remains to estimate $|a_k(f)| \cdot |S_{n,q}(F_k; \overline{G})(z) - F_k(z)|$, firstly for all $0 \leq k \leq n+q$ and secondly for $k \geq n+q+1$, where

$$S_{n,q}(F_k; \overline{G})(z) = \sum_{p=0}^{n+q} \binom{n+q}{p} [\Delta_{1/n}^p F_k(0)] \cdot F_p(z).$$

First it is useful to observe that by Gaier [38], p. 48, combined with Cauchy's theorem, for any fixed $1 < \beta < R$ we have

$$\mathcal{F}_k(w) := \frac{1}{2\pi i} \int_{|u|=\beta} \frac{F_k[\Psi(u)]}{u - w} du = w^k = e_k(w), \quad \text{for all } |w| < \beta.$$

Denote

$$D_{n,p,,q,k} = \binom{n+q}{p} \Delta_{1/n}^p e_k(0) = \binom{n+q}{p} [0, 1/n, \ldots, p/n; e_k] \cdot (p!)/n^p.$$

It follows

$$S_{n,q}(F_k; \overline{G})(z) = \sum_{p=0}^{n+q} D_{n,p,q,k} \cdot F_p(z).$$

Since e_k is convex of any order, it follows that all $D_{n,p,q,k} \geq 0$ and taking into account that by the Remark 1 just before the statement of Theorem 1.3.1,

$S_{n,p}(f)(1) = f(1 + p/n)$, we get $\sum_{p=0}^{n+q} D_{n,p,q,k} = \sum_{p=0}^{k} D_{n,p,q,k} = \frac{(n+q)^k}{n^k}$, for all k and n.

Also, note that $D_{n,k,q,k} = \frac{(n+q)(n+q-1)\ldots(n+q-k+1)}{n^k}$.

In the estimation of $|a_k(f)| \cdot |S_{n,q}(F_k; \overline{G})(z) - F_k(z)|$, we distinguish two cases: 1) $0 \leq k \leq n + q$; 2) $k > n + q$.

Case 1. We have

$$|S_{n,q}(F_k; \overline{G})(z) - F_k(z)| \leq |F_k(z)| \cdot |1 - D_{n,k,q,k}| + \sum_{p=0}^{k-1} D_{n,p,q,k} \cdot |F_p(z)|.$$

Fix now $\beta > q + 1$ and r with $1 < r(q+1) < \beta$.

By the inequality (8), p. 43 in Suetin [136], we have

$$|F_p(z)| \leq C(r)r^p, \text{ for all } z \in \overline{G}_r, p \geq 0.$$

Indeed, by relationship (8) above mentioned (with r instead of R there), we have $|F_p(z)| \leq C(r)r^p$, for all $z \in \Gamma_r$, which, by the Maximum Modulus Theorem for analytic functions, implies $|F_p(z)| \leq c(r)r^p$ for all $z \in \overline{G}_r$ (for these estimates, see also Curtiss [29], page 583, relationship (4.1) and the next two lines).

It is also worth noting that similar estimates hold from page 42, relationships (1), (3), and (4) in Suetin [136], by taking there $r = 1 + \varepsilon$ and $K = \overline{G}_{r'}$, with $1 < r' < r$ arbitrary close to r (in this case we get $|F_p(z)| \leq C(r)r^p$, for all $z \in \overline{G}_{r'}, p \geq 0$, but which still is good enough for the proof because r in $1 < r < R$ is arbitrary and $r' < r$ is arbitrary close to r).

As a consequence, we immediately get

$$|S_{n,q}(F_k; \overline{G})(z) - F_k(z)| \leq C(r)|1 - D_{n,k,q,k}|r^k + C(r)\left|\frac{(n+q)^k}{n^k} - D_{n,k,q,k}\right|r^k.$$

On the other hand, we immediately get

$$|1 - D_{n,k,q,k}| \leq \left|1 - \frac{(n+q)\ldots(n+q-k+1)}{(n+q)^k}\right|$$

$$+ (n+q)\ldots(n+q-k+1)\left[\frac{1}{n^k} - \frac{1}{(n+q)^k}\right]$$

$$= \left|1 - \frac{(n+q)\ldots(n+q-k+1)}{(n+q)^k}\right|$$

$$+ \frac{(n+q)\ldots(n+q-k+1)}{(n+q)^k}\left[\frac{(n+q)^k}{n^k} - 1\right].$$

But

$$\frac{(n+q)\dots(n+q-k+1)}{(n+q)^k}\left[\frac{(n+q)^k}{n^k}-1\right] \leq \left[\frac{(n+q)^k}{n^k}-1\right]$$

$$\leq \frac{(q+1)^k}{n}$$

and by using the inequality $1 - \Pi_{i=1}^k x_i \leq \sum_{i=1}^k (1-x_i)$ (valid if all $x_i \in [0,1]$), we get

$$1 - \frac{(n+q)\dots(n+q-k+1)}{(n+q)^k} = 1 - \Pi_{i=0}^{k-1}\frac{n+q-i}{n+q}$$

$$\leq \sum_{i=0}^{k-1}(1-(n+q-i)/(n+q)) = \frac{1}{n+q}\sum_{i=0}^{k-1}i = \frac{k(k-1)}{2(n+q)}.$$

Therefore, collecting the inequalities obtained, it follows

$$|1 - D_{n,k,q,k}| \leq \frac{k(k-1)}{2(n+q)} + \frac{(q+1)^k}{n} \leq \frac{k(k-1)}{n}(q+1)^k + \frac{(q+1)^k}{n}.$$

Also,

$$\left|\frac{(n+q)^k}{n^k} - D_{n,k,q,k}\right| = \frac{(n+q)^k}{n^k}\cdot\left|1 - \frac{(n+q)\dots(n+q-k+1)}{(n+q)^k}\right|$$

$$\leq \frac{(n+q)^k}{n^k}\cdot\frac{k(k-1)}{2(n+q)} \leq (q+1)^k\cdot\frac{k(k-1)}{2(n+q)},$$

which implies

$$|\mathcal{S}_{n,q}(F_k;\overline{G})(z) - F_k(z)| \leq \frac{2C(r)k(k-1)}{n}[(q+1)r]^k + \frac{C(r)}{n}[(q+1)r]^k$$

$$\leq \frac{3C(r)(k+1)(k+2)}{n}[(q+1)r]^k,$$

for all $z \in \overline{G}_r$.

Also, by the above formula for $a_k(f)$ we easily obtain $|a_k(f)| \leq \frac{C(\beta,f)}{\beta^k}$, for all $k \geq 0$. Note that $C(r), C(\beta,f) > 0$ are constants independent of k. For all $z \in \overline{G}_r$ and $k = 0,1,2,\dots n$, it follows

$$|a_k(f)|\cdot|\mathcal{S}_{n,q}(F_k;\overline{G})(z) - F_k(z)| \leq \frac{C(r,\beta,f)}{n}(k+1)(k+2)\left[\frac{(q+1)r}{\beta}\right]^k,$$

that is, for all $z \in \overline{G}_r$ we have

$$\sum_{k=0}^{n+q}|a_k(f)|\cdot|\mathcal{S}_{n,q}(F_k;\overline{G})(z) - F_k(z)| \leq \frac{C(r,\beta,f)}{n}\sum_{k=0}^{n+q}(k+1)(k+2)d^k,$$

where $0 < d = [(q+1)r]/\beta < 1$. Also, clearly we have $\sum_{k=0}^{n+q}(k+1)(k+2)d^k \le \sum_{k=0}^{\infty}(k+1)(k+2)d^k < \infty$, which finally implies that

$$\sum_{k=0}^{n+q}|a_k(f)| \cdot |S_{n,q}(F_k;\overline{G})(z) - F_k(z)| \le \frac{C^*(r,\beta,q,f)}{n}.$$

Case 2. We have

$$\sum_{k=n+q+1}^{\infty}|a_k(f)| \cdot |S_{n,q}(F_k;\overline{G})(z) - F_k(z)| \le \sum_{k=n+q+1}^{\infty}|a_k(f)| \cdot |S_{n,q}(F_k;\overline{G})(z)|$$

$$+ \sum_{k=n+q+1}^{\infty}|a_k(f)| \cdot |F_k(z)|.$$

By the estimates mentioned in Case 1, we immediately get

$$\sum_{k=n+q+1}^{\infty}|a_k(f)| \cdot |F_k(z)| \le C(r,\beta,q,f) \sum_{k=n+q+1}^{\infty} d^k, \text{ for all } z \in \overline{G}_r,$$

with $d = [(q+1)r]/\beta$.

Also,

$$\sum_{k=n+q+1}^{\infty}|a_k(f)| \cdot |S_{n,q}(F_k;\overline{G})(z)| = \sum_{k=n+q+1}^{\infty}|a_k(f)| \cdot \left|\sum_{p=0}^{n+q} D_{n,p,q,k} \cdot F_p(z)\right|$$

$$\le \sum_{k=n+q+1}^{\infty}|a_k(f)| \cdot \sum_{p=0}^{n+q} D_{n,p,q,k} \cdot |F_p(z)|.$$

But for $p \le n+q < k$ and taking into account the estimates obtained in Case 1, we get

$$|a_k(f)| \cdot |F_p(z)| \le C(r,\beta,q,f)\frac{r^p}{\beta^k} \le C(r,\beta,f)\frac{[(q+1)r]^k}{\beta^k}, \text{ for all } z \in \overline{G}_r,$$

which implies

$$\sum_{k=n+q+1}^{\infty}|a_k(f)| \cdot |S_{n,q}(F_k;\overline{G})(z) - F_k(z)|$$

$$\le C(r,\beta,q,f) \sum_{k=n+q+1}^{\infty}\sum_{p=0}^{n} D_{n,p,k}\left[\frac{(q+1)r}{\beta}\right]^k$$

$$= C(r, \beta, q, f) \sum_{k=n+q+1}^{\infty} \left[\frac{(q+1)r}{\beta} \right]^k$$

$$= C(r, \beta, q, f) \frac{d^{n+q+1}}{1-d},$$

with $d = [(q+1)r]/\beta$.

In conclusion, collecting the estimates in Cases 1 and 2, we obtain

$$|S_{n,q}(f; \overline{G})(z) - f(z)| \leq \frac{C_1}{n} + C_2 d^{n+q+1} \leq \frac{C}{n}, \; z \in \overline{G}_r, \; n \in \mathbb{N}.$$

This proves the theorem. □

1.4 Beta Operators of the First Kind

In this section, we study the approximation properties in the complex plane of the complex Beta operators of the first kind. In this sense, the exact order of simultaneous approximation and Voronovskaja-kind results with quantitative estimate for the complex Beta operators of the first kind attached to analytic functions in strips of compact disks are obtained. In this way, we put in evidence the overconvergence phenomenon for this operator, namely, the extensions of approximation properties with upper and exact quantitative estimates, from the real interval $(0, 1)$ to strips in compact disks of the complex plane of the form $SD^r(0, 1) = \{z \in \mathbb{C}; |z| \leq r, 0 < Re(z) < 1\}$ and $SD^r[a, b] = \{z \in \mathbb{C}; |z| \leq r, a \leq Re(z) \leq b\}$, with $r \geq 1$ and $0 < a < b < 1$.

The complex Beta operator of the first kind, firstly introduced in the case of real variable in Mühlbach [114] and studied in the real case by, e.g., Lupas [101], Khan [90], and Abel, Gupta, and Mohapatra [1], can be defined for all $n \in \mathbb{N}$ and $z \in \mathbb{C}$ satisfying $0 < Re(z) < 1$, by

$$K_n(f, z) = \frac{1}{B(nz, n(1-z))} \int_0^1 t^{nz-1}(1-t)^{n(1-z)-1} f(t) dt,$$

where $B(\alpha, \beta)$ is Euler's Beta function, defined by

$$B(\alpha, \beta) = \int_0^1 t^{\alpha-1}(1-t)^{\beta-1}, \alpha, \beta \in \mathbb{C}, \; Re(\alpha), \; Re(\beta) > 0.$$

In the sequel, we shall need the following auxiliary results.

Lemma 1.4.1 (Gal–Gupta [66]). *For all $e_p = t^p, p \in \mathbb{N} \cup \{0\}$, $n \in \mathbb{N}$, $z \in \mathbb{C}$ with $0 < Re(z) < 1$, we have $K_n(e_0, z) = 1$ and*

$$K_n(e_{p+1}, z) = \frac{nz + p}{n + p} K_n(e_p, z).$$

Proof. By the relationship of definition for the Beta operators, it is obvious that $K_n(e_0, z) = 1$. Next

$$K_n(e_{p+1}, z) = \frac{1}{B(nz, n(1-z))} B(nz + p + 1, n(1-z))$$

$$= \frac{1}{B(nz, n(1-z))} \left[(nz + p) \frac{\Gamma(nz + p)\Gamma(n(1-z))}{(n+p)\Gamma(n+p)} \right] = \frac{nz + p}{n + p} K_n(e_p, z).$$

This completes the proof of Lemma 1.4.1. $\qquad\square$

Lemma 1.4.2 (Gal–Gupta [66]). *If f is analytic in $D_R = \{z \in \mathbb{C}; |z| < R\}$, namely, $f(z) = \sum_{k=0}^{\infty} c_k z^k$, for all $z \in D_R$, then for all $n \in \mathbb{N}$, $1 \le r < R$ and $z \in SD^r(0, 1)$, we have*

$$K_n(f, z) = \sum_{k=0}^{\infty} c_k \cdot K_n(e_k, z).$$

Moreover, for any $0 < a < b < 1$, the convergence of the above series is uniform in $SD^r[a, b]$.

Proof. Let $|z| \le r$ and $0 < Re(z) < 1$. Defining $f_m(z) = \sum_{k=0}^{m} c_k e_k(z)$, by the linearity of K_n, it follows that $K_n(f_m, z) = \sum_{k=0}^{m} c_k K_n(e_k, z)$, for all $|z| \le r$, $0 < Re(z) < 1$ and $n, m \in \mathbb{N}$. It suffices to prove that for any fixed $n \in \mathbb{N}$, we have $\lim_{m \to \infty} K_n(f_m, z) = K_n(f, z)$, uniformly in any compact strip $SD^r[a, b]$ with $0 < a < b < 1$.

Firstly, we note that since from Andrews, Askey, and Roy [11], p. 8, formula (1.1.26), we have the following representation of the analytically continued Beta function

$$B(x, y) = \frac{\Gamma(x)\Gamma(y)}{\Gamma(x+y)} = \frac{x+y}{xy} \cdot \Pi_{n=1}^{\infty} \left[\frac{1 + (x+y)/n}{(1+x/n)(1+y/n)} \right], \quad Re(x), Re(y) > 0,$$

it follows that $B(x, y) = 0$ if and only if $x + y = 0$ or $x + y = -n$ with $n \in \mathbb{N}$. This implies that $B(nz, n(1-z)) \ne 0$ because $nz + n(1-z) = n > 0$.

Therefore, we get

$$|K_n(f_m, z) - K_n(f, z)|$$

$$= \frac{1}{|B(nz, n(1-z))|} \cdot \left| \int_0^1 t^{nz-1}(1-t)^{n(1-z)-1}[f_m(t) - f(t)]dt \right|$$

$$\le \frac{1}{|B(nz, n(1-z))|} \cdot \|f_m - f\|_r \int_0^1 |t^{nz-1}(1-t)^{n(1-z)-1}|dt,$$

where $|B(nz, n(1 - z))| > 0$ for $|z| \leq r$, $0 < Re(z) < 1$ and $n \in \mathbb{N}$. Now, by the continuity of $|B(nz, n(1 - z))|$ as function of z, it follows that for any $0 < a < b < 1$, there exist $A, M > 0$ both depending on a, b, r, such that $A \leq |B(nz, n(1 - z))| \leq M$, for all $z \in SD^r[a, b]$. This immediately implies that there exists a positive constant $C_{a,b,r,A,M}(f) > 0$ (independent of m), such that for all $z \in SD^r[a, b]$, we have

$$|K_n(f_m, z) - K_n(f, z)| \leq C_{a,b,r,A,M}(f)\|f_m - f\|_r, \ m \in \mathbb{N},$$

which for $m \to \infty$ proves the lemma. $\square$

In the first main result for $K_n(f)$, one refers to upper estimate.

Corollary 1.4.3 (Gal–Gupta [66]). *Let $R > 1$ and $f(z) = \sum_{k=0}^{\infty} c_k z^k$ for all $|z| < R$. Take $1 \leq r < R$. For all $z \in SD^r(0, 1)$ and $n \in \mathbb{N}$, we have*

$$|K_n(f, z) - f(z)| \leq \frac{C_r(f)}{n},$$

where $C_r(f) = \frac{1+r}{2} \cdot \sum_{k=2}^{\infty} |c_k| k(k - 1) r^{k-1} < \infty$.

Proof. Suppose that $|z| \leq r$ with $0 < Re(z) < 1$. By Lemma 1.4.2 we have $K_n(f, z) = \sum_{k=0}^{\infty} c_k K_n(e_k, z)$. Therefore we get

$$|K_n(f, z) - f(z)| \leq \sum_{k=0}^{\infty} |c_k| \cdot |K_n(e_k, z) - e_k(z)| = \sum_{k=2}^{\infty} |c_k| \cdot |K_n(e_k, z) - e_k(z)|,$$

as $K_n(e_0, z) = e_0(z) = 1$ and $K_n(e_1, z) = e_1(z) = z$.

By using now Lemma 1.4.1, for all $|z| \leq r$, $0 < Re(z) < 1$ and $n \in \mathbb{N}$, we get

$$|K_n(e_{k+1}, z) - e_{k+1}(z)|$$

$$= \left| \frac{nz + k}{n + k} K_n(e_k)(z) - \frac{nz + k}{n + k} e_k(z) + \frac{nz + k}{n + k} e_k(z) - e_{k+1}(z) \right|$$

$$\leq \frac{|nz + k|}{n + k} |K_n(e_k)(z) - e_k(z)| + |e_k(z)| \cdot \left| \frac{nz + k}{n + k} - z \right|$$

$$\leq r |K_n(e_k)(z) - e_k(z)| + r^k \frac{|k(1 - z)|}{n + k} \leq r |K_n(e_k)(z) - e_k(z)| + r^k (1 + r)\frac{k}{n},$$

for all $k = 0, 1, \ldots,$.

Taking above $k = 1, 2, \ldots$, step by step, we easily get by recurrence that

$$|K_n(e_k)(z) - e_k(z)| \leq r^{k-1}(1 + r)\frac{1}{n} [1 + 2 + \ldots + (k - 1)]$$

$$= r^{k-1}(1 + r)\frac{k(k - 1)}{2n},$$

for all $|z| \leq r$, $0 < Re(z) < 1$ and $n \in \mathbb{N}$, which immediately implies the corollary. $\qquad \square$

The following Voronovskaja-type result with a quantitative estimate holds.

Theorem 1.4.4 (Gal–Gupta [66]). *Let $R > 1$ and suppose that $f : D_R \to \mathbb{C}$ is analytic in $D_R = \{z \in \mathbb{C} : |z| < R\}$, that is, we can write $f(z) = \sum_{k=0}^{\infty} c_k z^k$, for all $z \in D_R$. For any fixed $r \in [1, R)$ and for all $|z| \leq r$ with $0 < Re(z) < 1$ and $n \in \mathbb{N}$, we have*

$$\left| K_n(f, z) - f(z) - \frac{z(1-z)f''(z)}{2n} \right| \leq \frac{M_r(f)}{n^2},$$

where $M_r(f) = \sum_{k=2}^{\infty} |c_k|(1+r)k(k+1)(k-1)^2 r^{k-1} < \infty$.

Proof. Denoting $\pi_{k,n}(z) = K_n(e_k)(z)$ and

$$E_{k,n}(z) = \pi_{k,n}(z) - e_k(z) - \frac{z^{k-1}(1-z)k(k-1)}{2n},$$

firstly it is clear that $E_{0,n}(z) = E_{1,n}(z) = 0$. Then, we can write

$$\left| K_n(f, z) - f(z) - \frac{z(1-z)f''(z)}{2n} \right| \leq \sum_{k=2}^{\infty} |c_k| \cdot |E_{k,n}(z)|,$$

so it remains to estimate $E_{k,n}(z)$ for $k \geq 2$.

In this sense, simple calculation based on Lemma 1.4.1 too leads us to the formula

$$E_{k,n}(z) = \frac{nz + k - 1}{n + k - 1} \cdot E_{k-1,n}(z) + \frac{z^{k-2}(1-z)(k-1)^2}{2n(n+k-1)} \cdot [k(1-z) - 2].$$

This immediately implies, for all $k \geq 2$ and $|z| \leq r$ with $0 < Re(z) < 1$,

$$|E_{k,n}(z)| \leq r|E_{k-1,n}(z)| + \frac{r^{k-2}(1+r)(k-1)^2}{2n^2} \cdot [k(1+r) + 2]$$

$$\leq r|E_{k-1,n}(z)| + \frac{r^{k-1}(1+r)(k-1)^2(k+1)}{n^2}.$$

Taking in the last inequality, $k = 2, 3, \ldots$, and reasoning by recurrence, finally we easily obtain

$$|E_{k,n}(z)| \leq \frac{r^{k-1}(1+r)}{n^2} \left[\sum_{j=1}^{k} (j-1)^2(j+1) \right] \leq \frac{r^{k-1}(1+r)}{n^2} \cdot k(k-1)^2(k+1).$$

We conclude that

$$\left| K_n(f, z) - f(z) - \frac{z(1-z)f''(z)}{2n} \right| \leq \sum_{k=2}^{\infty} |c_k| \cdot |E_{k,n}|$$

$$\leq \frac{1}{n^2} \sum_{k=2}^{\infty} |c_k|(1+r)k(k-1)^2(k+1)r^{k-1}.$$

As $f^{(4)}(z) = \sum_{k=4}^{\infty} c_k k(k-1)(k-2)(k-3)z^{k-4}$ and the series is absolutely convergent in $|z| \leq r$, it easily follows that $\sum_{k=4}^{\infty} |c_k|k(k-1)(k-2)(k-3)r^{k-4} < \infty$, which implies that $\sum_{k=2}^{\infty} |c_k|(1+r)k(k-1)^2(k+1)r^{k-1} < \infty$. This completes the proof of the theorem.   □

In what follows, we obtain the exact order in approximation by this type of complex Beta operators of the first kind and by their derivatives. In this sense, we present the following three results.

Theorem 1.4.5 (Gal–Gupta [66]). *Let $R > 1$ and suppose that $f : D_R \to \mathbb{C}$ is analytic in D_R, that is, we can write $f(z) = \sum_{k=0}^{\infty} c_k z^k$, for all $z \in D_R$. If f is not a polynomial of degree ≤ 1, then for any $r \in [1, R)$ and any $0 < a < b < 1$, we have*

$$\|K_n(f, \cdot) - f\|_{SD^r[a,b]} \geq \frac{C_{r,a,b}(f)}{n}, n \in \mathbb{N},$$

where $SD^r[a, b] = \{z \in \mathbb{C} : |z| \leq r, a \leq Re(z) \leq b\}$, $\|f\|_{SD^r[a,b]} = \sup\{|f(z)|; z \in SD^r[a, b]\}$ and $C_{r,a,b}(f)$ depends only on f, a, b, and r.

Proof. For all $|z| \leq r$ with $0 < Re(z) < 1$ and $n \in \mathbb{N}$, we have

$$K_n(f, z) - f(z) = \frac{1}{n}\left[\frac{z(1-z)f''(z)}{2} \right.$$

$$\left. + \frac{1}{n}\left\{ n^2\left(K_n(f, z) - f(z) - \frac{z(1-z)f''(z)}{2n} \right) \right\} \right].$$

Also, we have

$$\|F + G\|_{SD^r[a,b]} \geq \left| \|F\|_{SD^r[a,b]} - \|G\|_{SD^r[a,b]} \right| \geq \|F\|_{SD^r[a,b]} - \|G\|_{SD^r[a,b]}.$$

It follows

$$\|K_n(f, \cdot) - f\|_{SD^r[a,b]} \geq \frac{1}{n}\left[\left\| \frac{e_1(1-e_1)f''}{2} \right\|_{SD^r[a,b]} \right.$$

$$\left. - \frac{1}{n}\left\{ n^2 \left\| K_n(f, \cdot) - f - \frac{e_1(1-e_1)f''}{2n} \right\|_{SD^r[a,b]} \right\} \right].$$

Taking into account that by hypothesis f is not a polynomial of degree ≤ 1 in D_R, we get $\|e_1(1 - e_1)f''\|_{SD^r[a,b]} > 0$.

Indeed, supposing the contrary it follows that $z(1-z)f''(z) = 0$ for all $SD^r[a,b]$. Therefore we get $f''(z) = 0$, for all $z \in SD^r[a,b]$. Because f is analytic in D_R, by the uniqueness of analytic functions, we get $f''(z) = 0$, for all $z \in D_R$, that is, f is a linear function in D_R, which contradicts the hypothesis.

Now by Theorem 1.4.4, we have

$$n^2 \left\| K_n(f,\cdot) - f - \frac{e_1(1-e_1)f''}{2n} \right\|_{SD^r[a,b]} \leq M_r(f).$$

Therefore there exists an index n_0 depending only on f, a, b, and r, such that for all $n \geq n_0$, we have

$$\left\| \frac{e_1(1-e_1)f''}{2} \right\|_{SD^r[a,b]} - \frac{1}{n} \left\{ n^2 \left\| K_n(f,\cdot) - f - \frac{e_1(1-e_1)f''}{2n} \right\|_{SD^r[a,b]} \right\}$$

$$\geq \frac{1}{4} \| e_1(1-e_1)f'' \|_{SD^r[a,b]},$$

which immediately implies

$$\| K_n(f,\cdot) - f \|_{SD^r[a,b]} \geq \frac{1}{4n} \| e_1(1-e_1)f'' \|_{SD^r[a,b]}, \forall n \geq n_0.$$

For $n \in \{1, 2, \ldots n_0 - 1\}$ we obviously have $\| K_n(f,\cdot) - f \|_{SD^r[a,b]} \geq \frac{M_{r,a,b,n}(f)}{n}$ with $M_{r,a,b,n}(f) = n\| K_n(f,\cdot) - f \|_{SD^r[a,b]} > 0$. Indeed, if we would have $\| K_n(f,\cdot) - f \|_{SD^r[a,b]} = 0$, then $K_n(f,z) = f(z)$ would follow for all $z \in SD^r[a,b]$, which is valid only for f a linear function, contradicting the hypothesis on f. Therefore, finally we obtain $\| K_n(f,\cdot) - f \|_{SD^r[a,b]} \geq \frac{C_{r,a,b}(f)}{n}$ for all n, where

$$C_{r,a,b}(f) =$$

$$= \min\{ M_{r,a,b,1}(f), M_{r,a,b,2}(f) \ldots, M_{r,a,b,n_0-1}(f), \frac{1}{4} \| e_1(1-e_1)f'' \|_{SD^r[a,b]} \},$$

which completes the proof.   □

As a consequence of Corollary 1.4.3 and Theorem 1.4.5, we have the following:

Corollary 1.4.6 (Gal–Gupta [66]). *Let $R > 1$ and suppose that $f : D_R \to \mathbb{C}$ is analytic in D_R. If f is not a polynomial of degree ≤ 1, then for any $r \in [1, R)$ and any $0 < a < b < 1$, we have*

$$\| K_n(f,\cdot) - f \|_{SD^r[a,b]} \sim \frac{1}{n}, n \in \mathbb{N},$$

where the constants in the equivalence depend only on f, a, b, and r.

Our last result is in simultaneous approximation and can be stated as follows.

Theorem 1.4.7 (Gal–Gupta [66]). *Let $R > 1$ and suppose that $f : D_R \to \mathbb{C}$ is analytic in D_R, i.e., $f(z) = \sum_{k=0}^{\infty} c_k z^k$, for all $z \in D_R$, $1 \leq r < r_1 < R$, $0 < a_1 < a < b < b_1 < 1$ and $p \in \mathbb{N}$ be fixed. If f is not a polynomial of degree $\leq \max\{1, p-1\}$, then we have*

$$\|K_n^{(p)}(f, \cdot) - f^{(p)}\|_{SD^r[a,b]} \sim \frac{1}{n}, \, n \in \mathbb{N},$$

where the constants in the equivalence depend only on $f, r, r_1, a, b, a_1, b_1$, and p.

Proof. Denote by $\Gamma = \Gamma_{a_1, b_1, r_1} = S_1 \bigcup A_1 \bigcup S_2 \bigcup A_2$ the closed curve composed by the segments in $\mathbb{C}$

$$S_1 = \left\{ z = x + iy \in \mathbb{C}; x = a_1 \text{ and } -\sqrt{r_1^2 - a_1^2} \leq y \leq \sqrt{r_1^2 - a_1^2} \right\},$$

$$S_2 = \left\{ z = x + iy \in \mathbb{C}; x = b_1 \text{ and } -\sqrt{r_1^2 - b_1^2} \leq y \leq \sqrt{r_1^2 - b_1^2} \right\},$$

and by the arcs A_1, A_2 on the circle of center origin and radius r_1, situated in the region between the two segments defined above.

Clearly that Γ together with its interior is exactly $SD^{r_1}[a_1, b_1]$ and that from $r < r_1$ we have $SD^r[a, b] \subset SD^{r_1}[a_1, b_1]$, the inclusion beng strictly. By Cauchy's integral formula for derivatives, we have for all $z \in SD^r[a, b]$ and $n \in \mathbb{N}$

$$K_n^{(p)}(f, z) - f^{(p)}(z) = \frac{p!}{2\pi i} \int_{\Gamma} \frac{K_n(f, u) - f(u)}{(u - z)^{p+1}} du,$$

which, by Corollary 1.4.3 and by the inequality $|u - z| \geq d = \min\{r_1 - r, a - a_1, b_1 - b\}$ valid for all $z \in SD^r[a, b]$ and $u \in \Gamma$, implies

$$\|K_n^{(p)}(f, \cdot) - f^{(p)}\|_{SD^r[a,b]} \leq \frac{p!}{2\pi} \cdot \frac{l(\Gamma)}{d^{p+1}} \|K_n(f, \cdot) - f\|_{SD^{r_1}[a_1, b_1]}$$

$$\leq \frac{1}{n} \cdot \frac{C_{r_1}(f) p! l(\Gamma)}{2\pi d^{p+1}}.$$

Note that here, by simple geometrical reasonings, for the length $l(\Gamma)$ of Γ, we get

$$l(\Gamma) = l(S_1) + l(S_2) + l(A_1) + l(A_2)$$

$$= 2(\sqrt{r_1^2 - a_1^2} + \sqrt{r_1^2 - b_1^2}) + 2r_1[arccos(a_1/r_1) - arccos(b_1/r_1)],$$

where $arccos(\alpha)$ is considered expressed in radians.

It remains to prove the lower estimation for $\|K_n^{(p)}(f, \cdot) - f^{(p)}\|_{SD^r[a,b]}$.

By the proof of Theorem 1.4.5, for all $u \in \Gamma$ and $n \in \mathbb{N}$, we have

$$K_n(f, u) - f(u) = \frac{1}{n} \left[\frac{u(1-u)f''(u)}{2} \right.$$

$$\left. + \frac{1}{n} \left\{ n^2 \left(K_n(f, u) - f(u) - \frac{u(1-u)f''(u)}{2n} \right) \right\} \right].$$

Substituting it in the above Cauchy's integral formula, we get

$$K_n^{(p)}(f, z) - f^{(p)}(z) = \frac{1}{n} \left[\left(\frac{z(1-z)f''(z)}{2} \right)^{(p)} \right.$$

$$\left. + \frac{1}{n} \cdot \frac{p!}{2\pi i} \int_\Gamma \frac{n^2 \left(K_n(f, u) - f(u) - \frac{u(1-u)f''(u)}{2n} \right)}{(u-z)^{p+1}} du \right].$$

Thus

$$\left\| K_n^{(p)}(f, \cdot) - f^{(p)} \right\|_{SD^r[a,b]} \geq \frac{1}{n} \left[\left\| \left[\frac{e_1(1-e_1)f''}{2} \right]^{(p)} \right\|_{SD^r[a,b]} \right.$$

$$- \frac{1}{n} \left\| \frac{p!}{2\pi i} \int_\Gamma \frac{n^2 \left(K_n(f, u) - f(u) - \frac{u(1-u)f''(u)}{2n} \right)}{(u-\cdot)^{p+1}} du \right\|_{SD^r[a,b]} \right].$$

Applying Theorem 1.4.4 too, it follows

$$\left\| \frac{p!}{2\pi i} \int_\Gamma \frac{n^2 \left(K_n(f, u) - f(u) - \frac{u(1-u)f''(u)}{2n} \right)}{(u-\cdot)^{p+1}} du \right\|_{SD^r[a,b]}$$

$$\leq \frac{p!}{2\pi} \frac{l(\Gamma)n^2}{d^{p+1}} \left\| K_n(f, \cdot) - f - \frac{e_1(1-e_1)f''}{2n} \right\|_{SD^{r_1}[a_1,b_1]} \leq \frac{M_{r_1}(f)l(\Gamma)p!}{2\pi d^{p+1}}.$$

But by the hypothesis on f, we necessarily have

$$\| [e_1(1-e_1)f''/2]^{(p)} \|_{SD^r[a,b]} > 0.$$

Indeed, supposing the contrary, we get that $e_1(1-e_1)f''$ is a polynomial of degree $\leq p-1$ in $SD^r[a,b]$, which by the uniqueness of analytic functions implies that

$$z(1-z)f''(z) = Q_{p-1}(z) \text{ for all } z \in D_R,$$

where $Q_{p-1}(z)$ is a polynomial of degree $\leq p-1$.

Now, if $p = 1$ and $p = 2$, then the analyticity of f in D_R easily implies that f necessarily is a polynomial of degree $\leq 1 = \max\{1, p-1\}$. If $p > 2$, then the analyticity of f in D_R easily implies that f necessarily is a polynomial of degree $\leq p - 1 = \max\{1, p - 1\}$. Therefore, in all the cases we get a contradiction with the hypothesis.

In conclusion, $\|[e_1(1-e_1)f''/2]^{(p)}\|_{SD^r[a,b]} > 0$ and in continuation, reasoning exactly as in the proof of Theorem 1.4.5, but for $\|K_n^{(p)}(f, \cdot) - f^{(p)}\|_{SD^r[a,b]}$ instead of $\|K_n(f, \cdot) - f\|_{SD^r[a,b]}$, we immediately get the desired conclusion.

$\square$

1.5 Genuine Bernstein–Durrmeyer Polynomials

In this section, the order of simultaneous approximation and Voronovskaja-kind results with quantitative estimate for the complex genuine Durrmeyer polynomials attached to analytic functions on compact disks are obtained. In this way, we put in evidence the overconvergence phenomenon for the genuine Durrmeyer polynomials, namely, the extensions of the approximation properties (with quantitative estimates) from real intervals to compact disks in the complex plane.

More exactly, we obtain approximation results for the complex genuine Durrmeyer polynomials (which are different from the complex classical Bernstein–Durrmeyer operators), given explicitly and studied in the case of real variable in, e.g., Abel and Gupta [2], Chen [24], Goodman and Sharma [73], Parvanov [119], and Sauer [128] and some of its q-generalizations in Gupta [75], and Gupta and Finta [76], defined by

$$U_n(f)(z) = f(0)p_{n,0}(z) + f(1)p_{n,n}(z) + (n-1)\sum_{k=1}^{n-1} p_{n,k}(z) \int_0^1 p_{n-2,k-1}(t)f(t)dt,$$

where $p_{n,k}(z) = \binom{n}{k}z^k(1-z)^{n-k}$.

Firstly, let us note that from the formula of definition, it is easy to show that $U_n(e_0)(z) = 1$, $U_n(e_1)(z) = e_1(z)$, $U_n(e_2)(z) = e_2(z) + \frac{2z(1-z)}{n+1}$. Recall that we use the notations $e_p(z) = z^p$, $p \in \mathbb{N} \bigcup \{0\}$, $z \in \mathbb{C}$.

The following recurrence formula will be very useful.

Lemma 1.5.1 (Gal [50]). *For all $p \in \mathbb{N} \bigcup \{0\}$, $n \in \mathbb{N}$, and $z \in \mathbb{C}$, we have*

$$U_n(e_{p+1})(z) = \frac{z(1-z)}{n+p}U_n'(e_p)(z) + \frac{nz+p}{n+p}U_n(e_p)(z).$$

Proof. In the case of $p = 0$, the recurrence is immediate from $U_n(e_0)(z) = 1$ and $U_n(e_1)(z) = e_1(z)$. Therefore, let us suppose that $p \geq 1$.

Denoting for simplicity

$$I = \int_0^1 \binom{n-2}{k-1}t^{k+p-1}(1-t)^{n-k-1}dt = \binom{n-2}{k-1}B(k+p, n-k),$$

from the formula of definition, we can write

$$U_n(e_p)(z) = z^n + (n-1)\sum_{k=1}^{n-1} p_{n,k}(z) \cdot I,$$

which implies

$$U_n'(e_p)(z) = nz^{n-1} + (n-1)\sum_{k=1}^{n-1}\binom{n}{k}z^{k-1}(1-z)^{n-k} \cdot [kI]-$$

$$(n-1)\sum_{k=1}^{n-1}\binom{n}{k}z^k(1-z)^{n-k-1}\cdot[(n-k)I] = nz^{n-1} - \frac{n}{1-z}(n-1)\sum_{k=1}^{n-1} p_{n,k}(z)\cdot I+$$

$$\frac{1}{z(1-z)}\left[(n-1)\sum_{k=1}^{n-1} p_{n,k}(z)\cdot[kI]\right] = \frac{nz^n}{z} - \frac{n}{1-z}[U_n(e_p)(z) - z^n]+$$

$$\frac{1}{z(1-z)}\left[(n-1)\sum_{k=1}^{n-1} p_{n,k}(z)\cdot[kI]\right] =$$

$$\frac{1}{z(1-z)}\left[nz^n + (n-1)\sum_{k=1}^{n-1} p_{n,k}(z)\cdot[kI]\right] - \frac{n}{1-z}U_n(e_p)(z).$$

But by the formula for Euler's function $B(p,q)$ (see, e.g., Mocica [111], Exercise 1.31 b), p. 13) $B(p+1,q) = \frac{p}{p+q}B(p,q)$, for all $p,q \in \mathbb{R}$ with $p,q > 0$, we obtain

$$B(k+p+1, n-k) = \frac{k+p}{n+p}B(k+p, n-k),$$

which implies

$$kB(k+p, n-k) = (n+p)B(k+p+1, n-k) - pB(k+p, n-k).$$

Replacing above, we obtain
$$U_n'(e_p)(z) =$$

$$\frac{1}{z(1-z)}\left[nz^n + (n-1)\sum_{k=1}^{n-1} p_{n,k}(z)\binom{n-2}{k-1}[(n+p)B(k+p+1, n-k)-\right.$$

$$pB(k+p, n-k)]] - \frac{n}{1-z}U_n(e_p)(z) =$$

$$\frac{1}{z(1-z)}\left[(n+p)U_n(e_{p+1})(z) - pU_n(e_p)(z)\right] - \frac{n}{1-z}U_n(e_p)(z),$$

which by multiplication with $z(1-z)$ becomes the recurrence in the statement.
$\qquad\qquad\qquad\qquad\qquad\qquad\qquad\qquad\qquad\qquad\qquad\qquad\qquad$ $\square$

Also, the next lemma will be useful.

Lemma 1.5.2 (Gal [50]).

(i) For all $n \in \mathbb{N}$ and $p \in \mathbb{N} \bigcup \{0\}$, we have $U_n(e_p)(1) = 1$.

(ii) For all $n, p \in \mathbb{N}$ and $z \in \mathbb{C}$, we have

$$U_n(e_p)(z) =$$

$$\frac{(n-1)!}{(n-1+p)!} \sum_{k=0}^{n} \binom{n}{k} \Delta_1^k F_p(0) z^k = \frac{(n-1)!}{(n-1+p)!} \sum_{k=0}^{\min\{n,p\}} \binom{n}{k} \Delta_1^k F_p(0) z^k,$$

where $F_p(v) = \Pi_{j=0}^{p-1}(v+j)$, for all $v \geq 0$, and $\Delta_1^k F_p(0) = \sum_{j=0}^{k}(-1)^j \binom{k}{j} F_p(k-j)$ and $\Delta_1^k F_p(0) \geq 0$ for all k and p.

Proof. (i) From the relationship (obtained in the proof of Lemma 1.5.1)

$$U_n(e_p)(z) = z^n + (n-1) \sum_{k=1}^{n-1} p_{n,k}(z) \cdot I,$$

it is immediate that $U_n(e_p)(1) = 1$.

(ii) Since $U_n(e_0)(z) = 1$, we can suppose that $p \geq 1$. We have

$$U_n(e_p)(z) = z^n + (n-1) \sum_{k=1}^{n-1} p_{n,k}(z) \cdot I,$$

where $I = \binom{n-2}{k-1} B(k+p, n-k)$. Taking into account the formula $B(p,q) = \frac{(p-1)!(q-1)!}{(p+q-1)!}$, $p, q \in \mathbb{N}$, we easily obtain

$$I = \frac{(n-2)!}{(k-1)!(n-k-1)!} \cdot \frac{(k+p-1)!(n-k-1)!}{(n-1+p)!} =$$

$$\frac{(n-2)!}{(n-1+p)!}[k(k+1)\dots(k+p-1)] = \frac{(n-2)!}{(n-1+p)!} F_p(k),$$

where $F_p(v) = \Pi_{j=0}^{p-1}(v+j)$. It is clear here that $F_p(v)$ and its derivatives of any order are ≥ 0 for all $v \geq 0$, which imply that $\Delta_1^k F_p(0) \geq 0$ for all k and p.

Therefore we can write

$$U_n(e_p)(z) = z^n + \frac{(n-1)!}{(n-1+p)!} \sum_{k=1}^{n-1} p_{n,k}(z) \cdot F_p(k),$$

and by simple reasonings we easily obtain

$$U_n(e_p)(z) =$$

$$z^n + \frac{(n-1)!}{(n-1+p)!} \left[\sum_{k=0}^n p_{n,k}(z) \cdot F_p(k) - p_{n,0}(z)F_p(0) - p_{n,n}(z)F_p(n) \right] =$$

$$z^n + \frac{(n-1)!}{(n-1+p)!} \left[\sum_{k=0}^n \binom{n}{k} \Delta_1^k F_p(0) z^k - z^n [n(n+1)\ldots(n-1+p)] \right] =$$

$$\frac{(n-1)!}{(n-1+p)!} \sum_{k=0}^n \binom{n}{k} \Delta_1^k F_p(0) z^k = \frac{(n-1)!}{(n-1+p)!} \sum_{k=0}^{\min\{n,p\}} \binom{n}{k} \Delta_1^k F_p(0) z^k,$$

which proves the lemma. $\qquad\square$

The first main result is the following upper estimate.

Corollary 1.5.3 (Gal [50]). *Let $r \geq 1$:*

(i) For all $p, n \in \mathbb{N} \bigcup \{0\}$ and $|z| \leq r$ we have $|U_n(e_p)(z)| \leq r^p$.

(ii) Let $f(z) = \sum_{k=0}^\infty c_k z^k$ for all $|z| < R$ and take $1 \leq r < R$. For all $|z| \leq r$ and $n \in \mathbb{N}$, we have

$$|U_n(f)(z) - f(z)| \leq \frac{C_r(f)}{n},$$

where $C_r(f) = 2 \sum_{p=1}^\infty |c_p| p(p-1) r^p = 2 \sum_{p=2}^\infty |c_p| p(p-1) r^p < \infty$.

Proof. (i) By Lemma 1.5.2, (i) and (ii), it is immediate that

$$\frac{(n-1)!}{(n-1+p)!} \sum_{k=0}^n \binom{n}{k} \Delta_1^k F_p(0) = 1,$$

which implies

$$|U_n(e_p)(z)| \leq \frac{(n-1)!}{(n-1+p)!} \sum_{k=0}^n \binom{n}{k} \Delta_1^k F_p(0) r^k \leq$$

$$r^p \frac{(n-1)!}{(n-1+p)!} \sum_{k=0}^n \binom{n}{k} \Delta_1^k F_p(0) = r^p,$$

which proves (i).

(ii) First we prove that $U_n(f)(z) = \sum_{k=0}^\infty c_k U_n(e_k)(z)$. Indeed, denoting $f_m(z) = \sum_{j=0}^m c_j z^j$, $|z| \leq r$, $m \in \mathbb{N}$, since from the linearity of U_n, we obviously have $U_n(f_m)(z) = \sum_{k=0}^m c_k U_n(e_k)(z)$, it suffices to prove that for any fixed $n \in \mathbb{N}$ and $|z| \leq r$ with $r \geq 1$, we have $\lim_{m\to\infty} U_n(f_m)(z) = U_n(f)(z)$. But this is immediate from $\lim_{m\to\infty} \|f_m - f\|_r = 0$ (here $\|f\|_r = \max_{|z|\leq r}\{|f(z)|\}$) and from the inequality

$$|U_n(f_m)(z) - U_n(f)(z)| \le |f_m(0) - f(0)| \cdot |(1-z)^n| + |f_m(1) - f(1)| \cdot |z^n| +$$

$$(n-1) \sum_{k=1}^{n-1} |p_{n,k}(z)| \cdot \int_0^1 p_{n-2,k-1}(t)|f_m(t) - f(t)| dt \le C_{r,n} \|f_m - f\|_r,$$

valid for all $|z| \le r$, where simple calculation gives

$$C_{r,n} = (1+r)^n + r^n + (n-1) \sum_{k=1}^{n-1} \binom{n}{k} (1+r)^{n-k} r^k \cdot \int_0^1 p_{n-2,k-1}(t) dt.$$

Therefore we get

$$|U_n(f)(z) - f(z)| \le \sum_{p=0}^{\infty} |c_p| \cdot |U_n(e_p)(z) - e_p(z)| = \sum_{p=1}^{\infty} |c_p| \cdot |U_n(e_p)(z) - e_p(z)|,$$

since $U_n(e_0)(z) = e_0(z) = 1$.

We have two cases: 1) $1 \le p \le n$; 2) $p > n$.

Case 1. From Lemma 1.5.2 (i) and (ii), we obtain

$$U_n(e_p)(z) - e_p(z) =$$

$$z^p \left[\frac{(n-1)!}{(n-1+p)!} \binom{n}{p} \Delta_1^p F_p(0) - 1 \right] + \frac{(n-1)!}{(n-1+p)!} \sum_{k=0}^{p-1} \binom{n}{k} \Delta_1^k F_p(0) z^k,$$

and

$$|U_n(e_p)(z) - e_p(z)| \le r^p \left[1 - \frac{(n-1)!}{(n-1+p)!} \binom{n}{p} \Delta_1^p F_p(0) \right] +$$

$$r^p \left[1 - \frac{(n-1)!}{(n-1+p)!} \binom{n}{p} \Delta_1^p F_p(0) \right] \le 2r^p \left[1 - \frac{(n-1)!}{(n-1+p)!} \binom{n}{p} \Delta_1^p F_p(0) \right].$$

Here it is easy to see that we can write

$$\frac{(n-1)!}{(n-1+p)!} \binom{n}{p} \Delta_1^p F_p(0) = \frac{(n-1)!}{(n-1+p)!} \binom{n}{p} p! = \Pi_{j=1}^p \frac{n+j-p}{n+j-1}.$$

But applying the formula (easily proved by mathematical induction)

$$1 - \Pi_{j=1}^k x_j \le \sum_{j=1}^k (1 - x_j), 0 \le x_j \le 1, j = 1, \dots, k,$$

for $x_j = \frac{n+j-p}{n+j-1}$ and $k = p$, we obtain

$$1 - \Pi_{j=1}^{p} \frac{n+j-p}{n+j-1} \leq \sum_{j=1}^{p} \left(1 - \frac{n+j-p}{n+j-1}\right) =$$

$$(p-1) \sum_{j=1}^{p} \frac{1}{n+j-1} \leq \frac{(p-1)p}{n}.$$

Therefore it follows

$$|U_n(e_p)(z) - e_p(z)| \leq \frac{2p(p-1)r^p}{n}.$$

Case 2. By (i) and by $p > n \geq 1$, we obtain

$$|U_n(e_p)(z) - e_p(z)| \leq |U_n(e_p)(z)| + |e_p(z)| \leq 2r^p < \frac{2p}{n}r^p \leq \frac{2p(p-1)}{n}r^p.$$

In conclusion, from both Cases 1 and 2, we obtain for all $p, n \in \mathbb{N}$

$$|U_n(e_p)(z) - e_p(z)| \leq \frac{2p(p-1)r^p}{n},$$

which implies

$$|U_n(f)(z) - f(z)| \leq \frac{2}{n} \sum_{p=1}^{\infty} |c_p| p(p-1) r^p$$

and proves the corollary. $\square$

The Voronovskaja's theorem for the real case with a quantitative estimate is obtained by Proposition 7.4 in Gonska, Pitul, and Rasa [71] in the following form:

$$\left| U_n(f)(x) - f(x) - \frac{x(1-x)}{n+1} f''(x) \right| \leq \frac{x(1-x)}{n+1} \omega_1 \left(f'' \frac{2}{3\sqrt{n+3}} \right),$$

$n \in \mathbb{N}, x \in [0,1]$.

For the complex genuine Durrmeyer polynomials it is expected to derive a formula of the form

$$\left| U_n(f)(z) - f(z) - \frac{z(1-z)}{n+1} f''(z) \right| \leq \frac{M_{r,f}}{n^2},$$

for all $n \in \mathbb{N}, |z| \leq r$.

Indeed, in what follows we will prove the Voronovskaja theorem with a quantitative estimate for the complex version of genuine Durrmeyer polynomials.

Theorem 1.5.4 (Gal [50]). *Let $R > 1$ and suppose that $f : \mathbb{D}_R \to \mathbb{C}$ is analytic in $\mathbb{D}_R = \{z \in \mathbb{C}; |z| < R\}$, that is, we can write $f(z) = \sum_{k=0}^{\infty} c_k z^k$, for all $z \in \mathbb{D}_R$.*

For any fixed $r \in [1, R)$ and for all $n \in \mathbb{N}, |z| \leq r$, the following Voronovskaja-type result holds:

$$\left| U_n(f)(z) - f(z) - \frac{z(1-z)f''(z)}{n+1} \right| \leq \frac{M_r(f)}{n^2},$$

where $M_r(f) = 16 \sum_{k=3}^{\infty} |c_k| k(k-1)(k-2)^2 r^k < \infty$.

Proof. Denoting $\pi_{k,n}(z) = U_n(e_k)(z)$, by the proof of Corollary 1.5.3 (ii), we can write $U_n(f)(z) = \sum_{k=0}^{\infty} c_k \pi_{k,n}(z)$. Also, since

$$\frac{z(1-z)f''(z)}{n+1} = \frac{z(1-z)}{n+1} \sum_{k=2}^{\infty} k(k-1)c_k z^{k-2},$$

taking into account that $U_n(e_0)(z) = 1$, $U_n(e_1)(z) = e_1(z)$, we immediately obtain

$$\left| U_n(f)(z) - f(z) - \frac{z(1-z)f''(z)}{n+1} \right| \leq$$

$$\sum_{k=2}^{\infty} |c_k| \cdot \left| \pi_{k,n}(z) - e_k(z) - \frac{k(k-1)(1-z)z^{k-1}}{n+1} \right|,$$

for all $z \in \overline{\mathbb{D}}_R$, $n \in \mathbb{N}$.

In what follows, we will use the recurrence obtained in Lemma 1.5.1:

$$\pi_{k+1,n}(z) = \frac{z(1-z)}{n+k} \pi'_{k,n}(z) + \frac{nz+k}{n+k} \pi_{k,n}(z),$$

for all $n \in \mathbb{N}$, $z \in \mathbb{C}$, and $k = 0, 1, \dots$.

If we denote

$$E_{k,n}(z) = \pi_{k,n}(z) - e_k(z) - \frac{k(k-1)(1-z)z^{k-1}}{n+1},$$

then it is clear that $E_{k,n}(z)$ is a polynomial of degree $\leq k$, and by a simple calculation and the use of the above recurrence, we obtain the following relationship:

$$E_{k,n}(z) = \frac{z(1-z)}{n+k-1} E'_{k-1,n}(z) + \frac{nz+k-1}{n+k-1} E_{k-1,n}(z) +$$

$$\frac{z^{k-2}(1-z)(k-1)(k-2)}{(n+1)(n+k-1)}[(2k-3) - 2kz],$$

valid for all $k \geq 2$, $n \in \mathbb{N}$, and $|z| \leq r$.

For all $k, n \in \mathbb{N}$, $k \geq 2$ and $|z| \leq r$, it implies

$$|E_{k,n}(z)| \leq \frac{r(1+r)}{n+k-1}|E'_{k-1,n}(z)| + \frac{nr+k-1}{n+k-1}|E_{k-1,n}(z)|+$$

$$\frac{r^{k-2}(1+r)(k-1)(k-2)}{(n+1)(n+k-1)}[2k-3+2kr].$$

Since $\frac{r(1+r)}{n+k+1} \leq \frac{r(1+r)}{n}$ and $\frac{nr+k-1}{n+k-1} \leq r$, it follows

$$|E_{k,n}(z)| \leq \frac{r(1+r)}{n}|E'_{k-1,n}(z)| + r|E_{k-1,n}(z)|+$$

$$\frac{r^{k-2}(1+r)^2 2k(k-1)(k-2)}{n^2}.$$

Now we will estimate $|E'_{k-1,n}(z)|$, for $k \geq 2$. We will use the estimate obtained in the proof of Corollary 1.5.3 (ii),

$$|\pi_{k,n}(z) - e_k(z)| \leq \frac{2k(k-1)r^k}{n},$$

for all $k, n \in \mathbb{N}$, $|z| \leq r$, with $1 \leq r$.

Taking into account that $E_{k-1,n}(z)$ is a polynomial of degree $\leq (k-1)$, by the well-known Bernstein's inequality, we obtain

$$|E'_{k-1,n}(z)| \leq \frac{k-1}{r}\|E_{k-1,n}\|_r \leq$$

$$\frac{k-1}{r}\left[\|\pi_{k-1,n} - e_{k-1}\|_r + \left\|\frac{(k-1)(k-2)e_{k-2}(1-e_1)}{n}\right\|_r\right] \leq$$

$$\frac{k-1}{r}\left[\frac{2(k-1)(k-2)r^{k-1}}{n} + \frac{r^{k-2}(r+1)(k-1)(k-2)}{n}\right] \leq$$

$$\frac{k-1}{r} \cdot \frac{(k-1)(k-2)r^{k-2}}{n}[2r+(r+1)] \leq \frac{4(k-1)^2(k-2)r^{k-2}}{n},$$

for all $k \geq 2$ and $|z| \leq r$.

This implies

$$\frac{r(1+r)}{n}|E'_{k-1,n}(z)| \leq \frac{4(k-1)^2(k-2)(1+r)r^{k-1}}{n^2},$$

and

$$|E_{k,n}(z)| \leq r|E_{k-1,n}(z)| + \frac{4(k-1)^2(k-2)(1+r)r^{k-1}}{n^2}+$$

$$\frac{r^{k-2}(1+r)^2 2k(k-1)(k-2)}{n^2} \leq r|E_{k-1,n}(z)| + \frac{8(k-1)^2(k-2)r^k}{n^2}+$$

$$\frac{8r^k k(k-1)(k-2)}{n^2} \leq r|E_{k-1,n}(z)| + \frac{16k(k-1)(k-2)r^k}{n^2}.$$

But $E_{0,n}(z) = E_{1,n}(z) = E_{2,n}(z) = 0$, for any $z \in \mathbb{C}$, and therefore by writing the last inequality for $k = 3, 4, \ldots$, we easily obtain, step by step, the following:

$$|E_{k,n}(z)| \leq \frac{16 r^k}{n^2} \left[\sum_{j=3}^{k} j(j-1)(j-2) \right] \leq \frac{16 r^k k(k-1)(k-2)^2}{n^2}.$$

As a conclusion, we obtain

$$\left| U_n(f)(z) - f(z) - \frac{z(1-z)f''(z)}{n+1} \right| \leq \sum_{k=3}^{\infty} |c_k| \cdot |E_{k,n}(z)| \leq$$

$$\frac{16}{n^2} \sum_{k=3}^{\infty} |c_k| k(k-1)(k-2)^2 r^k.$$

Note that since $f^{(4)}(z) = \sum_{k=5}^{\infty} c_k k(k-1)(k-2)(k-3)z^{k-4}$ and the series is absolutely convergent in $|z| \leq r$, it easily follows that $\sum_{k=5}^{\infty} |c_k| k(k-1)(k-2)(k-3)r^{k-4} < \infty$, which immediately implies that $\sum_{k=3}^{\infty} |c_k| k(k-1)(k-2)^2 r^k < \infty$ and proves the theorem. $\qquad \square$

By using the above Voronovskaja's theorem, in what follows we will obtain the exact order in approximation by the complex genuine Durrmeyer polynomials and their derivatives. In this sense, we present the following results.

Theorem 1.5.5 (Gal [50]). *Let $R > 1$, $\mathbb{D}_R = \{z \in \mathbb{C}; |z| < R\}$ and let us suppose that $f : \mathbb{D}_R \to \mathbb{C}$ is analytic in $\mathbb{D}_R$, that is, we can write $f(z) = \sum_{k=0}^{\infty} c_k z^k$, for all $z \in \mathbb{D}_R$. If f is not a polynomial of degree ≤ 1, then for any $r \in [1, R)$, we have*

$$\|U_n(f) - f\|_r \geq \frac{C_r(f)}{n+1}, n \in \mathbb{N},$$

where the constant $C_r(f)$ depends only on f and r.

Proof. For all $z \in \mathbb{D}_R$ and $n \in \mathbb{N}$ we have

$$U_n(f)(z) - f(z) = \frac{1}{n+1} \{z(1-z)f''(z) +$$

$$\frac{1}{n+1} \left[(n+1)^2 \left(U_n(f)(z) - f(z) - \frac{z(1-z)f''(z)}{n+1} \right) \right] \}.$$

In what follows we will apply to this identity the following obvious property:

$$\|F + G\|_r \geq |\|F\|_r - \|G\|_r| \geq \|F\|_r - \|G\|_r.$$

It follows

$$\|U_n(f) - f\|_r \geq$$

$$\frac{1}{n+1}\left\{\|e_1(1-e_1)f''\|_r - \frac{1}{n+1}\left[(n+1)^2\left\|U_n(f) - f - \frac{e_1(1-e_1)f''}{n+1}\right\|_r\right]\right\}.$$

Taking into account that by hypothesis f is not a polynomial of degree ≤ 1 in $\mathbb{D}_R$, we get $\|e_1(1-e_1)f''\|_r > 0$. Indeed, supposing the contrary it follows that $z(1-z)f''(z) = 0$ for all $z \in \overline{\mathbb{D}}_r$, which by the analyticity of f in the disk $|z| \leq r$ (with $r \geq 1$) clearly implies $f''(z) = 0$ for all $|z| \leq r$, that is, f is a polynomial of degree ≤ 1 for all $|z| \leq r$, a contradiction.

But by Theorem 1.5.4 we immediately get

$$(n+1)^2\left\|U_n(f) - f - \frac{e_1(1-e_1)f''}{n+1}\right\|_r \leq 4M_r(f).$$

Therefore, there exists an index n_0 depending only on f and r, such that for all $n \geq n_0$, we have

$$\|e_1(1-e_1)f''\|_r - \frac{1}{n+1}\left[(n+1)^2\left\|U_n(f) - f - \frac{e_1(1-e_1)f''}{n+1}\right\|_r\right] \geq$$

$$\frac{1}{2}\|e_1(1-e_1)f''\|_r,$$

which immediately implies

$$\|U_n(f) - f\|_r \geq \frac{1}{n+1}\cdot\frac{1}{2}\|e_1(1-e_1)f''\|_r, \forall n \geq n_0.$$

For $n \in \{1,\ldots,n_0-1\}$ we obviously have $\|U_n(f) - f\|_r \geq \frac{M_{r,n}(f)}{n+1}$ with $M_{r,n}(f) = (n+1)\cdot\|U_n(f) - f\|_r > 0$. Indeed, if we would have that $\|U_n(f) - f\|_r = 0$, then $U_n(f)(z) = f(z)$ would follow for all $|z| \leq r$, which is valid only for f a polynomial of degree ≤ 1, contradicting the hypothesis on f in the statement. Therefore, finally we get $\|U_n(f) - f\|_r \geq \frac{C_r(f)}{n+1}$ for all n, where

$$C_r(f) = \min\{M_{r,1}(f),\ldots,M_{r,n_0-1}(f), \frac{1}{2}\|e_1(1-e_1)f''\|_r\}.$$

This completes the proof. $\qquad\square$

Combining now Theorem 1.5.5 with Corollary 1.5.3, (ii), we immediately get the following.

Corollary 1.5.6 (Gal [50]). *Let $R > 1$, $\mathbb{D}_R = \{z \in \mathbb{C}; |z| < R\}$ and let us suppose that $f : \mathbb{D}_R \to \mathbb{C}$ is analytic in $\mathbb{D}_R$. If f is not a polynomial of degree ≤ 1, then for any $r \in [1, R)$, we have*

$$\|U_n(f) - f\|_r \sim \frac{1}{n}, n \in \mathbb{N},$$

where the constants in the equivalence depend only on f and r.

For the derivatives of complex genuine Durrmeyer polynomials, we can state the following result.

Theorem 1.5.7 (Gal [50]). *Let $\mathbb{D}_R = \{z \in \mathbb{C}; |z| < R\}$ be with $R > 1$ and let us suppose that $f : \mathbb{D}_R \to \mathbb{C}$ is analytic in $\mathbb{D}_R$, i.e., $f(z) = \sum_{k=0}^{\infty} c_k z^k$, for all $z \in \mathbb{D}_R$. Also, let $1 \le r < r_1 < R$ and $p \in \mathbb{N}$ be fixed. If f is not a polynomial of degree $\le \max\{1, p-1\}$, then we have*

$$\|U_n^{(p)}(f) - f^{(p)}\|_r \sim \frac{1}{n},$$

where the constants in the equivalence depend only on f, r, r_1, and p.

Proof. Denoting by Γ the circle of radius $r_1 >$ and center 0 (where $r_1 > r \ge 1$), by Cauchy's formulas it follows that for all $|z| \le r$ and $n \in \mathbb{N}$ we have

$$U_n^{(p)}(f)(z) - f^{(p)}(z) = \frac{p!}{2\pi i} \int_\Gamma \frac{U_n(f)(v) - f(v)}{(v-z)^{p+1}} dv,$$

which, by Corollary 1.5.3 (ii) and by the inequality $|v - z| \ge r_1 - r$ valid for all $|z| \le r$ and $v \in \Gamma$, immediately implies

$$\|U_n^{(p)}(f) - f^{(p)}\|_r \le \frac{p!}{2\pi} \cdot \frac{2\pi r_1}{(r_1 - r)^{p+1}} \|U_n(f) - f\|_{r_1} \le C_{r_1}(f) \frac{p! r_1}{n(r_1 - r)^{p+1}}.$$

It remains to prove the lower estimate for $\|U_n^{(p)}(f) - f^{(p)}\|_r$.

For this purpose, as in the proof of Theorem 1.5.5, for all $v \in \Gamma$ and $n \in \mathbb{N}$, we have

$$U_n(f)(v) - f(v) = \frac{1}{n+1} \{v(1-v)f''(v) +$$

$$\frac{1}{n+1} \left[(n+1)^2 \left(U_n(f)(v) - f(v) - \frac{v(1-v)f''(v)}{n+1} \right) \right] \},$$

which replaced in the above Cauchy's formula implies

$$U_n^{(p)}(f)(z) - f^{(p)}(z) = \frac{1}{n+1} \left\{ \frac{p!}{2\pi i} \int_\Gamma \frac{v(1-v)f''(v)}{(v-z)^{p+1}} dv + \right.$$

$$\left. \frac{1}{n+1} \cdot \frac{p!}{2\pi i} \int_\Gamma \frac{(n+1)^2 \left(U_n(f)(v) - f(v) - \frac{v(1-v)f''(v)}{n+1} \right)}{(v-z)^{p+1}} dv \right\} =$$

$$\frac{1}{n+1}\left\{[z(1-z)f''(z)]^{(p)}+\right.$$

$$\left.\frac{1}{n+1}\cdot\frac{p!}{2\pi i}\int_\Gamma\frac{(n+1)^2\left(U_n(f)(v)-f(v)-\frac{v(1-v)f''(v)}{n+1}\right)}{(v-z)^{p+1}}dv\right\}.$$

Passing now to $\|\cdot\|_r$, it follows

$$\|U_n^{(p)}(f)-f^{(p)}\|_r\geq\frac{1}{n+1}\left\{\left\|[e_1(1-e_1)f'']^{(p)}\right\|_r-\right.$$

$$\left.\frac{1}{n+1}\left\|\frac{p!}{2\pi}\int_\Gamma\frac{(n+1)^2\left(U_n(f)(v)-f(v)-\frac{v(1-v)f''(v)}{n+1}\right)}{(v-z)^{p+1}}dv\right\|_r\right\},$$

where by using Theorem 1.5.4 we get

$$\left\|\frac{p!}{2\pi}\int_\Gamma\frac{(n+1)^2\left(U_n(f)(v)-f(v)-\frac{v(1-v)f''(v)}{n+1}\right)}{(v-z)^{p+1}}dv\right\|_r\leq$$

$$\frac{p!}{2\pi}\cdot\frac{2\pi r_1(n+1)^2}{(r_1-r)^{p+1}}\left\|U_n(f)-f-\frac{e_1(1-e_1)f''}{n+1}\right\|_{r_1}\leq\frac{4M_{r_1}(f)p!r_1}{(r_1-r)^{p+1}}.$$

But by hypothesis on f, we have $\left\|[e_1(1-e_1)f'']^{(p)}\right\|_r>0.$

Indeed, supposing the contrary, it follows that $z(1-z)f''(z)$ is a polynomial of degree $\leq p-1$.

Now, if $p=1$, then we get $z(1-z)f''(z)=C$, which implies $f''(z)=\frac{C}{z(1-z)}$, for all $|z|\leq r$ with $r\geq 1$. But since $f''(z)$ is analytic in $|z|\leq r$, this necessarily implies $C=0$, that is, $f(z)$ is a polynomial of degree $\leq 1=\max\{1,p-1\}$, a contradiction with the hypothesis on f.

For $p=2$ we get $z(1-z)f''(z)=Az+B$, which implies $f''(z)=\frac{Az+B}{z(1-z)}$, for all $|z|\leq r$. But since $f''(z)$ is analytic in $|z|\leq r$, this necessarily implies $A=B=0$ (because contrariwise f would have a pole at $z=0$ or at $z=1$), and therefore f would be a polynomial of degree $\leq 1=\max\{1,p-1\}$, a contradiction.

If $p\geq 3$, then we get $z(1-z)f''(z)=Q_{p-1}(z)$, where $Q_{p-1}(z)$ is a polynomial of degree $\leq p-1$. This implies $f''(z)=\frac{Q_{p-1}(z)}{z(1-z)}$, for all $|z|\leq r$ (with $r\geq 1$). Then the analyticity of f obviously implies that $Q_{p-1}(z)=z(1-z)R_{p-3}(z)$ where $R_{p-3}(z)$ is a polynomial of degree $\leq p-3$ (because contrariwise f would have a pole at $z=0$ or at $z=1$). Therefore, necessarily we get $f''(z)=R_{p-3}(z)$, that is, $f(z)$ is a polynomial of degree $\leq p-1=\max\{1,p-1\}$, which again contradicts the hypothesis on f.

In continuation reasoning exactly as in the proof of Theorem 1.5.5, we immediately get the desired conclusion. $\qquad\square$

1.6 Bernstein–Durrmeyer Polynomials with Jacobi Weights

In this section, the exact order of simultaneous approximation and Voronovskaja-kind results with quantitative estimate for the complex Durrmeyer polynomials based on Jacobi weights and attached to analytic functions in compact disks are obtained. In this way, we put in evidence the overconvergence phenomenon for the Durrmeyer polynomials based on Jacobi weights, namely, the extensions of approximation properties (with quantitative estimates) from real intervals to compact disks in the complex plane.

More exactly, we obtain approximation results for the complex Durrmeyer polynomials based on the Jacobi weights (introduced and studied in the case of real variable in Pǎltǎnea [120–122], Berens and Xu [15], and Abel and Heilman [3]) defined for all $z \in \mathbb{C}$ by

$$M_n^{a,b}(f)(z) = \sum_{k=0}^{n} \frac{\binom{n}{k} z^k (1-z)^{n-k}}{B(a+k+1, b+n-k+1)} \int_0^1 f(t) t^{a+k} (1-t)^{b+n-k} dt,$$

where $a, b > -1$ and $B(a, b)$ is the Beta function.

For the proofs of our main results, first the following recurrence formula is needed.

Lemma 1.6.1 (Gal [51]). *For all $p, n \in \mathbb{N} \bigcup \{0\}$, $a, b > -1$ and $z \in \mathbb{C}$, we have*

$$M_n^{a,b}(e_{p+1})(z) =$$

$$\frac{z(1-z)}{n+p+a+b+2}[M_n^{a,b}(e_p)]'(z) + \frac{nz+p+a+1}{n+p+a+b+2} M_n^{a,b}(e_p)(z).$$

Proof. By simple calculation we obtain

$$[M_n^{a,b}(e_p)]'(z) = \sum_{k=0}^{n} \frac{\binom{n}{k} k z^{k-1} (1-z)^{n-k}}{B(a+k+1, b+n-k+1)} \int_0^1 t^{p+a+k} (1-t)^{n+b-k} dt -$$

$$\sum_{k=0}^{n} \frac{\binom{n}{k}(n-k) z^k (1-z)^{n-k-1}}{B(a+k+1, b+n-k+1)} \int_0^1 t^{p+a+k} (1-t)^{n+b-k} dt =$$

$$\frac{1}{z(1-z)} \left[\sum_{k=0}^{n} \frac{\binom{n}{k} z^k (1-z)^{n-k}}{B(a+k+1, b+n-k+1)} \int_0^1 k t^{p+a+k} (1-t)^{n+b-k} dt \right]$$

$$- \frac{n}{1-z} M_n^{a,b}(e_p)(z). \tag{1.6.1}$$

Denoting Euler's Beta function $B(p, q) = \int_0^1 t^{p-1}(1-t)^{q-1} dt$, it is known that we have the formula (see, e.g., Mocica [111], Exercise 1.31 b), p. 13) $(p+q)B(p+1, q) = pB(p, q)$, for all $p, q \in \mathbb{R}$ with $p, q \geq 0$.

This allows to immediately prove the relationship

$$(n + p + a + b + 2)B(p + a + k + 2, n + b - k + 1) =$$

$$(p + a + k + 1)B(p + a + k + 1, n + b - k + 1),$$

which is equivalent to

$$(n + p + a + b + 2) \int_0^1 t^{p+a+k+1}(1 - t)^{n+b-k}dt =$$

$$(p + a + k + 1) \int_0^1 t^{p+a+k}(1 - t)^{n+b-k}dt,$$

and to

$$\int_0^1 kt^{p+a+k}(1 - t)^{n+b-k}dt = (n + p + a + b + 2) \int_0^1 t^{p+a+k+1}(1 - t)^{n+b-k}dt -$$

$$(p + a + 1) \int_0^1 t^{p+a+k}(1 - t)^{n+b-k}dt.$$

Using this last formula in (1.6.1), we obtain

$$[M_n^{a,b}(e_p)]'(z) = \frac{1}{z(1 - z)}(n + p + a + b + 2)M_n(e_{p+1})(z) -$$

$$\frac{p + a + 1}{z(1 - z)}M_n^{a,b}(e_p)(z) - \frac{n}{1 - z}M_n^{a,b}(e_p)(z),$$

which implies the recurrence in the statement.   □

Also, we need the following properties of $M_n^{a,b}(e_p)(z)$.

Lemma 1.6.2 (Gal [51]). *Let $a, b \in \mathbb{R}$ with $a, b > -1$. We have:*

(i) For all $n, p \in \mathbb{N}$ and $z \in \mathbb{C}$,

$$M_n^{a,b}(e_p)(z) = \sum_{k=0}^{\min\{n,p\}} \binom{n}{k} \Delta_1^k F_p(0) z^k,$$

where $F_p(v) = \Pi_{j=1}^p(v + a + j)/\Pi_{j=1}^p(n + a + b + j + 1)$ for all $v \geq 0$ and $\Delta_1^k F_p(0) \geq 0$ for all k and p.

(ii) $M_n^{a,b}(e_0)(z) = 1$.

(iii) $M_n^{a,b}(e_p)(1) = \sum_{k=0}^{\min\{n,p\}} \binom{n}{k} \Delta_1^k F_p(0) = \Pi_{j=1}^p \frac{a+n+j}{n+a+b+j+1} < 1$.

Proof. (i) Denote as usual $p_{n,k}(z) = \binom{n}{k} z^k (1 - z)^{n-k}$. By the formula of definition we immediately get

$$M_n^{a,b}(e_p)(z) = \sum_{k=0}^{n} p_{n,k}(z) \frac{B(p+a+k+1, b+n-k+1)}{B(a+k+1, b+n-k+1)}.$$

Applying p-times the formula (see, e.g., Mocica [111]) $B(p+1,q) = \frac{p}{p+q}B(p,q)$, by mathematical induction we easily obtain

$$\frac{B(p+a+k+1, b+n-k+1)}{B(a+k+1, b+n-k+1)} = \Pi_{j=1}^{p} \frac{(a+k+j)}{(n+a+b+j+1)} := A_{k,p}(a,b).$$

$$(1.6.2)$$

Therefore we get

$$M_n^{a,b}(e_p)(z) = \sum_{k=0}^{n} \binom{n}{k} A_{k,p}(a,b) z^k (1-z)^{n-k}.$$

Defining now $F_p(v) = \Pi_{j=1}^{p}(v+a+j)/\Pi_{j=1}^{p}(n+a+b+j+1)$, for all $v \geq 0$, by simple reasonings we easily can write

$$M_n^{a,b}(e_p)(z) = \sum_{k=0}^{n} \binom{n}{k} z^k (1-z)^{n-k} F_p(k) = \sum_{k=0}^{n} \binom{n}{k} \Delta_1^k F_p(0) z^k.$$

Here $F_p(v)$ is convex of any order on $[0, \infty)$ since it is easy to show that F_p and the derivatives of any order of F_p are ≥ 0 on $[0, \infty)$. This clearly implies $\Delta_1^k F_p(0) \geq 0$ for all k and p. Also, since $F_p(v)$ is a polynomial of degree p, we get $\Delta_1^k F_p(0) = 0$ for $k > p$, which implies

$$M_n^{a,b}(e_p)(z) = \sum_{k=0}^{\min\{n,p\}} \binom{n}{k} \Delta_1^k F_p(0) z^k.$$

(ii) By the formula of definition we get

$$M_n^{a,b}(e_0)(z) = \sum_{k=0}^{n} \frac{\binom{n}{k} z^k (1-z)^{n-k}}{B(a+k+1, b+n-k+1)} \int_0^1 t^{a+k}(1-t)^{b+n-k} dt =$$

$$\sum_{k=0}^{n} \frac{\binom{n}{k} z^k (1-z)^{n-k}}{B(a+k+1, b+n-k+1)} B(a+k+1, b+n-k+1) =$$

$$\sum_{k=0}^{n} \binom{n}{k} z^k (1-z)^{n-k} = 1.$$

(iii) By the formula of definition and using (1.6.2) for $k = n$ we immediately get

$$M_n^{a,b}(e_p)(1) =$$

$$\frac{\int_0^1 t^{p+a+n}(1-t)^b dt}{B(a+n+1,b+1)} = \frac{B(p+a+n+1,b+1)}{B(a+n+1,b+1)} =$$

$$\Pi_{j=1}^p \frac{a+n+j}{n+a+b+j+1} < 1.$$

Also, by (i) it follows that

$$M_n^{a,b}(e_p)(1) = \sum_{k=0}^{\min\{n,p\}} \binom{n}{k} \Delta_1^k F_p(0),$$

which proves the lemma. $\qquad\square$

In the first main result, one refers to upper estimates.

Corollary 1.6.3 (Gal [51]). *Let $a,b > -1$ and $r \geq 1$.*

(i) For all $p,n \in \mathbb{N}\bigcup\{0\}$ and $|z| \leq r$, we have $|M_n^{a,b}(e_p)(z)| \leq r^p$.
(ii) Let $f(z) = \sum_{k=0}^\infty c_k z^k$ for all $|z| < R$ and take $1 \leq r < R$. For all $|z| \leq r$ and $n \in \mathbb{N}$, we have

$$|M_n^{a,b}(f)(z) - f(z)| \leq \frac{C_{r,a,b}(f)}{n},$$

where $C_{r,a,b}(f) = 2 \sum_{p=1}^\infty |c_p| p(p+a+b+1) r^p < \infty$.

Proof. (i) By Lemma 1.6.2, (i), we obtain

$$|M_n^{a,b}(e_p)(z)| \leq \sum_{k=0}^{\min\{n,p\}} \binom{n}{k} \Delta_1^k F_p(0) r^k.$$

If $p \leq n$, then we get

$$|M_n^{a,b}(e_p)(z)| \leq \sum_{k=0}^p \binom{n}{k} \Delta_1^k F_p(0) r^k \leq r^p \sum_{k=0}^p \binom{n}{k} \Delta_1^k F_p(0) \leq r^p,$$

by taking into account Lemma 1.6.2 (iii).
If $p > n$, then

$$|M_n^{a,b}(e_p)(z)| \leq \sum_{k=0}^n \binom{n}{k} \Delta_1^k F_p(0) r^k \leq r^n < r^p,$$

which proves (i).

(ii) First we prove that $M_n^{a,b}(f)(z) = \sum_{k=0}^{\infty} c_k M_n^{a,b}(e_k)(z)$. Indeed, denoting $f_m(z) = \sum_{j=0}^{m} c_j z^j$, $|z| \leq r$, $m \in \mathbb{N}$, since from the linearity of $M_n^{a,b}$, we obviously have $M_n^{a,b}(f_m)(z) = \sum_{k=0}^{m} c_k M_n^{a,b}(e_k)(z)$, it suffices to prove that for any fixed $n \in \mathbb{N}$ and $|z| \leq r$ with $r \geq 1$, we have $\lim_{m \to \infty} M_n^{a,b}(f_m)(z) = M_n^{a,b}(f)(z)$. But this is immediate from $\lim_{m \to \infty} \|f_m - f\|_r = 0$ (here $\|f\|_r = \max_{|z| \leq r}\{|f(z)|\}$) and from the inequality

$$|M_n^{a,b}(f_m)(z) - M_n^{a,b}(f)(z)| \leq$$

$$\sum_{k=0}^{n} \frac{\binom{n}{k}|z^k(1-z)^{n-k}|}{B(a+k+1,b+n-k+1)} \int_0^1 |f_m(t) - f(t)| t^{a+k}(1-t)^{b+n-k} dt \leq$$

$$\sum_{k=0}^{n} \binom{n}{k}|z^k(1-z)^{n-k}| \cdot \|f_m - f\|_r \leq C_{r,n}\|f_m - f\|_r,$$

valid for all $|z| \leq r$.

Therefore we get

$$|M_n^{a,b}(f)(z) - f(z)| \leq \sum_{p=0}^{\infty} |c_p| \cdot |M_n^{a,b}(e_p)(z) - e_p(z)| =$$

$$\sum_{p=1}^{\infty} |c_p| \cdot |M_n^{a,b}(e_p)(z) - e_p(z)|,$$

since $M_n^{a,b}(e_0) = e_0$.

We have two cases: 1) $1 \leq p \leq n$; 2) $p > n$.

Case 1. From Lemma 1.6.2 (i) and (iii), we obtain

$$M_n^{a,b}(e_p)(z) - e_p(z) = z^p \left[\binom{n}{p} \Delta_1^p F_p(0) - 1 \right] + \sum_{k=0}^{p-1} \binom{n}{k} \Delta_1^k F_p(0) z^k,$$

and

$$|M_n^{a,b}(e_p)(z) - e_p(z)| \leq r^p \left[1 - \binom{n}{p} \Delta_1^p F_p(0) \right] +$$

$$r^p \left[\Pi_{j=1}^{p} \frac{a+n+j}{n+a+b+j+1} - \binom{n}{p} \Delta_1^p F_p(0) \right] \leq 2r^p \left[1 - \binom{n}{p} \Delta_1^p F_p(0) \right].$$

Here it is easy to see that we can write

$$\binom{n}{p} \Delta_1^p F_p(0) = \binom{n}{p} \frac{p!}{\Pi_{j=1}^{p}(n+a+b+j+1)} = \Pi_{j=1}^{p} \frac{n-p+j}{n+a+b+j+1}.$$

But applying the formula (easily proved by mathematical induction)

$$1 - \Pi_{j=1}^{k} x_j \leq \sum_{j=1}^{k} (1 - x_j), 0 \leq x_j \leq 1, j = 1, \ldots, k,$$

for $x_j = \frac{n-p+j}{n+a+b+j+1}$ and $k = p$, we obtain

$$1 - \Pi_{j=1}^{p} \frac{n-p+j}{n+a+b+j+1} \leq \sum_{j=1}^{p} \left(1 - \frac{n-p+j}{n+a+b+j+1}\right) =$$

$$(p+a+b+1)\sum_{j=1}^{p} \frac{1}{n+a+b+j+1} \leq \frac{p(p+a+b+1)}{n+a+b+2}.$$

Therefore it follows

$$|M_n^{a,b}(e_p)(z) - e_p(z)| \leq \frac{2p(p+a+b+1)r^p}{n+a+b+2} \leq \frac{2p(p+a+b+1)r^p}{n}.$$

Case 2. By (i) and by $p > n \geq 0$, we obtain

$$|M_n^{a,b}(e_p)(z) - e_p(z)| \leq$$

$$|M_n^{a,b}(e_p)(z)| + |e_p(z)| \leq 2r^p < \frac{2p}{n} r^p \leq \frac{2p(p+a+b+1)}{n} r^p.$$

In conclusion, from both Cases 1 and 2, we obtain for all $p, n \in \mathbb{N}$

$$|M_n^{a,b}(e_p)(z) - e_p(z)| \leq \frac{2p(p+a+b+1)r^p}{n},$$

which implies

$$|M_n^{a,b}(f)(z) - f(z)| \leq \frac{2}{n} \sum_{p=1}^{\infty} |c_p| p(p+a+b+1)r^p$$

and proves the corollary.   □

Remark. Voronovskaja's theorem in the real case is

$$\lim_{n \to \infty} \left\{ M_n^{a,b}(f)(x) - f(x) - \frac{1}{n} \cdot \frac{[x^{a+1}(1-x)^{b+1}f'(x)]'}{x^a(1-x)^b} \right\} = 0.$$

This suggests in the complex case a quantitative estimate of the form

$$\left| M_n^{a,b}(f)(z) - f(z) - \frac{1}{n} \cdot \frac{[z^{a+1}(1-z)^{b+1}f'(z)]'}{z^a(1-z)^b} \right| \leq \frac{C_{r,a,b}(f)}{n^2},$$

for all $n \in \mathbb{N}$, $|z| \leq r$.

In this sense, we present the following.

Theorem 1.6.4 (Gal [51]). *Let $R > 1$, $a, b > -1$ and suppose that $f :$ $\mathbb{D}_R \to \mathbb{C}$ is analytic in $\mathbb{D}_R$, that is, we can write $f(z) = \sum_{k=0}^{\infty} c_k z^k$, for all $z \in \mathbb{D}_R$. For any fixed $r \in [1, R)$ and for all $n \in \mathbb{N}, |z| \leq r$, the following Voronovskaja-type result holds:*

$$\left| M_n^{a,b}(f)(z) - f(z) - \frac{1}{n} \cdot \frac{[z^{a+1}(1-z)^{b+1}f'(z)]'}{z^a(1-z)^b} \right| \leq \frac{M_{r,a,b}(f)}{n^2},$$

where $M_{r,a,b}(f) = \sum_{k=1}^{\infty} |c_k| k B_{k,r,a,b} r^{k-1} < \infty$,

$$B_{k,r,a,b} = A_{k,r,a,b} + 4|k + a + b|(k-1)^2(1+r),$$

$$A_{k,r,a,b} = (k-1)|k-1+a| \cdot |2k - 2 + a| + r|Q_{3,a,b}(k)| + r^2|R_{3,a,b}(k)|,$$

$$Q_{3,a,b}(k) = -(k-1)(k-2)(k-1+a)-(k-1)^2(k+a+b)-(k-1)(k+a)(k+a+b)-$$
$$k^2(k+1+a+b) - ak(k+1+a+b),$$

and

$$R_{3,a,b}(k) = (k-1)^2(k+a+b) + k(k+1+a+b).$$

Proof. Denoting $e_k(z) = z^k$, $k = 0, 1, \ldots$, and $\pi_{k,n,a,b}(z) = M_n^{a,b}(e_k)(z)$, by the proof of Corollary 1.6.3, (ii), we can write $M_n^{a,b}(f)(z) = \sum_{k=0}^{\infty} c_k \pi_{k,n,a,b}(z)$.

Also, since

$$\frac{[z^{a+1}(1-z)^{b+1}f'(z)]'}{n} = \left(\frac{z^{a+1}(1-z)^{b+1}}{n} \sum_{k=1}^{\infty} k c_k z^{k-1} \right)'$$

$$= \left(\sum_{k=1}^{\infty} \frac{k c_k}{n} z^{k+a}(1-z)^{b+1} \right)'$$

$$= \sum_{k=1}^{\infty} \frac{k c_k}{n}[(k+a)z^{k+a-1}(1-z)^{b+1} - (b+1)z^{k+a}(1-z)^b]$$

$$= \sum_{k=1}^{\infty} c_k \frac{k z^{k+a-1}(1-z)^b}{n}[k + a - (k+a+b+1)z],$$

this immediately implies

$$\left| M_n^{a,b}(f)(z) - f(z) - \frac{1}{n} \cdot \frac{[z^{a+1}(1-z)^{b+1}f'(z)]'}{z^a(1-z)^b} \right|$$

$$\leq \sum_{k=1}^{\infty} |c_k| \cdot \left| \pi_{k,n,a,b}(z) - e_k(z) - \frac{k z^{k-1}[k + a - (k+a+b+1)z]}{n} \right|,$$

for all $z \in \overline{\mathbb{D}}_1$, $n \in \mathbb{N}$.

In what follows, we will use the recurrence obtained in Lemma 1.6.1

$$\pi_{k+1,n,a,b}(z) = \frac{z(1-z)}{n+k+a+b+2}\pi'_{k,n,a,b}(z) + \frac{nz+k+a+1}{n+k+a+b+2}\pi_{k,n,a,b}(z),$$

for all $n \in \mathbb{N}$, $z \in \mathbb{C}$ and $k = 0, 1, \ldots$.
  If we denote

$$E_{k,n,a,b}(z) = \pi_{k,n,a,b}(z) - e_k(z) - \frac{kz^{k-1}[k+a-(k+a+b+1)z]}{n},$$

then it is clear that $E_{k,n,a,b}(z)$ is a polynomial of degree $\leq k$, and by a simple calculation and the use of the above recurrence, we obtain the following relationship

$$E_{k,n,a,b}(z) = \frac{z(1-z)}{n+k+a+b+1}E'_{k-1,n,a,b}(z) + \frac{nz+a+k}{n+k+a+b+1}E_{k-1,n,a,b}(z)$$

$$+X_{k,n,a,b}(z),$$

where

$$X_{k,n,a,b}(z) =$$

$$\frac{x^{k-2}}{n(n+k+1+a+b)}[(k-1)(k-1+a)(2k-2+a)+zQ_{3,a,b}(k)+z^2R_{3,a,b}(k)],$$

with $Q_{3,a,b}(k)$ and $R_{3,a,b}(k)$ polynomials of degree ≤ 3 with respect to k, given by

$$Q_{3,a,b}(k) = -(k-1)(k-2)(k-1+a)-(k-1)^2(k+a+b)-(k-1)(k+a)(k+a+b)-$$

$$k^2(k+1+a+b) - ak(k+1+a+b),$$

and

$$R_{3,a,b}(k) = (k-1)^2(k+a+b) + k(k+1+a+b),$$

for all $k \geq 1$, $n \in \mathbb{N}$ and $|z| \leq r$.
  We will use the estimate obtained in the proof of Corollary 1.6.3, (ii),

$$|\pi_{k,n,a,b}(z) - e_k(z)| \leq \frac{2k(k+a+b+1)r^k}{n},$$

for all $k, n \in \mathbb{N}$, $|z| \leq r$, with $1 \leq r$.
  For all $k, n \in \mathbb{N}$, $k \geq 1$ and $|z| \leq r$, it implies

$$|E_{k,n,a,b}(z)| \leq$$

$$\frac{r(1+r)}{n+k+a+b+1}|E'_{k-1,n,a,b}(z)| + \frac{nr+a+k}{n+k+1}|E_{k-1,n,a,b}(z)| + |X_{k,n,a,b}(z)|.$$

Since $\frac{r(1+r)}{n+k+a+b+1} \leq \frac{r(1+r)}{n}$ and $\frac{nr+a+k}{n+k+a+b+1} \leq r$ for all $k \geq 1$, it follows

$$|E_{k,n,a,b}(z)| \leq \frac{r(1+r)}{n}|E'_{k-1,n,a,b}(z)| + r|E_{k-1,n,a,b}(z)| + |X_{k,n,a,b}(z)|.$$

Now we will estimate $|E'_{k-1,n,a,b}(z)|$, for $k \geq 1$. Taking into account that $E_{k-1,n,a,b}(z)$ is a polynomial of degree $\leq (k-1)$, we obtain

$$|E'_{k-1,n,a,b}(z)| \leq \frac{k-1}{r}\|E_{k-1,n,a,b}(z)\|_r \leq$$

$$\frac{k-1}{r}\left[\|\pi_{k-1,n,a,b} - e_{k-1}\|_r + \left\|\frac{(k-1)e_{k-2}[k+a-1-(k+a+b)e_1]}{n}\right\|_r\right] \leq$$

$$\frac{k-1}{r}\left[\frac{2(k-1)|k+a+b|r^{k-1}}{n} + \frac{r^{k-2}(k-1)(r+1)|k+a+b|}{n}\right] \leq$$

$$\frac{|k+a+b|(k-1)^2}{n}\left[2r^{k-2} + \frac{r+1}{r}r^{k-2}\right] \leq \frac{4|k+a+b|(k-1)^2 r^{k-2}}{n}.$$

This implies

$$\frac{r(1+r)}{n}|E'_{k-1,n,a,b}(z)| \leq \frac{4|k+a+b|(k-1)^2(1+r)r^{k-1}}{n^2},$$

and

$$|E_{k,n,a,b}(z)| \leq r|E_{k-1,n,a,b}(z)| + \frac{4|k+a+b|(k-1)^2(1+r)r^{k-1}}{n^2} + |X_{k,n,a,b}(z)|,$$

where

$$|X_{k,n,a,b}(z)| \leq \frac{r^{k-2}}{n^2}[(k-1)|k-1+a|\cdot|2k-2+a|+r|Q_{3,a,b}(k)|+r^2|R_{3,a,b}(k)|] \leq$$

$$\frac{r^{k-1}}{n^2}A_{k,r,a,b},$$

where

$$A_{k,r,a,b} = (k-1)|k-1+a|\cdot|2k-2+a| + r|Q_{3,a,b}(k)| + r^2|R_{3,a,b}(k)|,$$

for all $|z| \leq r$, $k \geq 1$, $n \in \mathbb{N}$.

Denoting now

$$B_{k,r,a,b} = A_{k,r,a,b} + 4|k+a+b|(k-1)^2(1+r),$$

we obtain

$$|E_{k,n,a,b}(z)| \leq r|E_{k-1,n,a,b}(z)| + \frac{r^{k-1}}{n^2}B_{k,r,a,b},$$

for all $|z| \leq r$, $k \geq 1$, $n \in \mathbb{N}$.

But $E_{0,n,a,b}(z) = 0$, for any $z \in \mathbb{C}$, and therefore by writing the last inequality for $k = 1, 2, \ldots$, we easily obtain, step by step, the following:

$$|E_{k,n,a,b}(z)| \leq \frac{r^{k-1}}{n^2} \left[\sum_{j=1}^{k} B_{j,r,a,b} \right] \leq \frac{kr^{k-1}}{n^2} B_{k,r,a,b}.$$

As a conclusion, we obtain

$$\left| M_n^{a,b}(f)(z) - f(z) - \frac{[z^{a+1}(1-z)^{b+1}f'(z)]'}{nz^a(1-z)^b} \right| \leq \sum_{k=1}^{\infty} |c_k| \cdot |E_{k,n,a,b}(z)| \leq$$

$$\frac{1}{n^2} \sum_{k=1}^{\infty} |c_k| k B_{k,r,a,b} r^{k-1}.$$

Note that since $f^{(4)}(z) = \sum_{k=4}^{\infty} c_k k(k-1)(k-2)(k-3)z^{k-4}$ and the series is absolutely convergent in $|z| \leq r$, it easily follows that $\sum_{k=4}^{\infty} |c_k| k(k-1)(k-2)(k-3)r^{k-4} < \infty$, which immediately implies that $\sum_{k=1}^{\infty} |c_k| k B_{k,r} r^{k-1} < \infty$ and proves the theorem. $\qquad \square$

Finally, we will obtain the exact order in approximation by complex modified Durrmeyer polynomials (with Jacobi weights) and their derivatives. In this sense we present the following results.

Theorem 1.6.5 (Gal [51]). *Let $R > 1$, $a, b > -1$, $\mathbb{D}_R = \{z \in \mathbb{C}; |z| < R\}$ and let us suppose that $f : \mathbb{D}_R \to \mathbb{C}$ is analytic in $\mathbb{D}_R$, that is, we can write $f(z) = \sum_{k=0}^{\infty} c_k z^k$, for all $z \in \mathbb{D}_R$. If f is not a polynomial of degree 0, then for any $r \in [1, R)$, we have*

$$\|M_n^{a,b}(f) - f\|_r \geq \frac{C_{r,a,b}(f)}{n}, n \in \mathbb{N},$$

where the constant $C_{r,a,b}(f)$ depends only on f, a, b, and r.

Proof. For all $z \in \mathbb{D}_R$ and $n \in \mathbb{N}$, we have

$$M_n^{a,b}(f)(z) - f(z) = \frac{1}{n} \left\{ \frac{[z^{a+1}(1-z)^{b+1}f'(z)]'}{z^a(1-z)^b} + \right.$$

$$\left. \frac{1}{n} \left[n^2 \left(M_n^{a,b}(f)(z) - f(z) - \frac{[z^{a+1}(1-z)^{b+1}f'(z)]'}{nz^a(1-z)^b} \right) \right] \right\}.$$

In what follows we will apply to this identity the following obvious property:

$$\|F + G\|_r \geq |\,\|F\|_r - \|G\|_r\,| \geq \|F\|_r - \|G\|_r.$$

It follows

$$\|M_n^{a,b}(f) - f\|_r \geq \frac{1}{n} \left\{ \left\| \frac{[e_1^{a+1}(1-e_1)^{b+1}f']'}{e_1^a(1-e_1)^b} \right\|_r - \right.$$

$$\left. \frac{1}{n} \left[n^2 \left\| M_n^{a,b}(f)(z) - f - \frac{[e_1^{a+1}(1-e_1)^{b+1}f']'}{ne_1^a(1-e_1)^b} \right\|_r \right] \right\}.$$

Taking into account that by hypothesis f is not a polynomial of degree 0 in $\mathbb{D}_R$, we get $\left\| \frac{[e_1^{a+1}(1-e_1)^{b+1}f']'}{e_1^a(1-e_1)^b} \right\|_r > 0$.

Indeed, supposing the contrary, it follows that $[z^{a+1}(1-z)^{b+1}f'(z)]' = 0$ for all $z \in \overline{\mathbb{D}}_r$, which implies $z^{a+1}(1-z)^{b+1}f'(z) = C$ for all $|z| \leq r$, that is, $f'(z) = \frac{C}{z^{a+1}(1-z)^{b+1}}$ for all $|z| \leq r$. But since f is analytic in $\overline{\mathbb{D}}_r$ (with $r \geq 1$), we necessarily have $C = 0$ (since contrariwise f would have poles at $z = 0$ and $z = 1$), which implies $f'(z) = 0$ and $f(z) = c$ for all $z \in \overline{\mathbb{D}}_r$, a contradiction.

Now, by Theorem 1.6.4 we have

$$n^2 \left\| M_n^{a,b}(f) - f - \frac{[e_1^{a+1}(1-e_1)^{b+1}f']'}{ne_1^a(1-e_1)^b} \right\|_r \leq M_{r,a,b}(f).$$

Therefore, there exists an index n_0 depending only on f and r, such that for all $n \geq n_0$, we have

$$\left\| \frac{[e_1^{a+1}(1-e_1)^{b+1}f']'}{e_1^a(1-e_1)^b} \right\|_r - \frac{1}{n} \left[n^2 \left\| M_n^{a,b}(f) - f - \frac{[e_1^{a+1}(1-e_1)^{b+1}f']'}{ne_1^a(1-e_1)^b} \right\|_r \right] \geq$$

$$\frac{1}{2} \left\| \frac{[e_1^{a+1}(1-e_1)^{b+1}f']'}{e_1^a(1-e_1)^b} \right\|_r,$$

which immediately implies

$$\|M_n^{a,b}(f) - f\|_r \geq \frac{1}{n} \cdot \frac{1}{2} \left\| \frac{[e_1^{a+1}(1-e_1)^{b+1}f']'}{e_1^a(1-e_1)^b} \right\|_r, \forall n \geq n_0.$$

For $n \in \{1, \ldots, n_0 - 1\}$ we obviously have $\|M_n^{a,b}(f) - f\|_r \geq \frac{M_{r,n,a,b}(f)}{n}$ with $M_{r,n,a,b}(f) = n \cdot \|M_n^{a,b}(f) - f\|_r > 0$, which finally implies $\|M_n^{a,b}(f) - f\|_r \geq \frac{C_{r,a,b}(f)}{n}$ for all n, where

$$C_{r,a,b}(f) = \min\{M_{r,1,a,b}(f), \ldots, M_{r,n_0-1,a,b}(f), \frac{1}{2} \left\| \frac{[e_1^{a+1}(1-e_1)^{b+1}f']'}{e_1^a(1-e_1)^b} \right\|_r \}.$$

This completes the proof. $\square$

Combining now Theorem 1.6.5 with Corollary 1.6.3 (ii), we immediately get the following.

Corollary 1.6.6 (Gal [51]). *Let $R > 1$, $a, b > -1$, $\mathbb{D}_R = \{z \in \mathbb{C}; |z| < R\}$ and let us suppose that $f : \mathbb{D}_R \to \mathbb{C}$ is analytic in $\mathbb{D}_R$. If f is not a polynomial of degree 0, then for any $r \in [1, R)$ we have*

$$\|M_n^{a,b}(f) - f\|_r \sim \frac{1}{n}, n \in \mathbb{N},$$

where the constants in the equivalence depend only on f, a, b, and r.

For the derivatives of complex modified Durrmeyer polynomials with Jacobi weights we can state the following result.

Theorem 1.6.7 (Gal [51]). *Let $\mathbb{D}_R = \{z \in \mathbb{C}; |z| < R\}$ be with $R > 1$, $a, b > -1$ and let us suppose that $f : \mathbb{D}_R \to \mathbb{C}$ is analytic in $\mathbb{D}_R$, i.e., $f(z) = \sum_{k=0}^{\infty} c_k z^k$, for all $z \in \mathbb{D}_R$. Also, let $1 \leq r < r_1 < R$ and $p \in \mathbb{N}$ be fixed. If f is not a polynomial of degree $\leq p - 1$, then we have*

$$\|[M_n^{a,b}(f)]^{(p)} - f^{(p)}\|_r \sim \frac{1}{n},$$

where the constants in the equivalence depend only on f, a, b, r, r_1, and p.

Proof. Denoting by Γ the circle of radius $r_1 >$ and center 0 (where $r_1 > r \geq 1$), by Cauchy's formulas it follows that for all $|z| \leq r$ and $n \in \mathbb{N}$, we have

$$[M_n^{a,b}(f)(z)]^{(p)} - f^{(p)}(z) = \frac{p!}{2\pi i} \int_{\Gamma} \frac{M_n^{a,b}(f)(v) - f(v)}{(v - z)^{p+1}} dv,$$

which, by Corollary 1.6.3 (ii) and by the inequality $|v - z| \geq r_1 - r$ valid for all $|z| \leq r$ and $v \in \Gamma$, immediately implies

$$\|[M_n^{a,b}(f)]^{(p)} - f^{(p)}\|_r \leq$$

$$\frac{p!}{2\pi} \cdot \frac{2\pi r_1}{(r_1 - r)^{p+1}} \|M_n^{a,b}(f) - f\|_{r_1} \leq M_{r_1, a, b}(f) \frac{p! r_1}{n(r_1 - r)^{p+1}}.$$

It remains to prove the lower estimate for $\|[M_n^{a,b}(f)]^{(p)} - f^{(p)}\|_r$.

For this purpose, as in the proof of Theorem 1.6.5, for all $v \in \Gamma$ and $n \in \mathbb{N}$, we have

$$M_n^{a,b}(f)(v) - f(v) = \frac{1}{n} \left\{ \frac{[v^{a+1}(1-v)^{b+1} f'(v)]'}{v^a (1-v)^b} + \right.$$

$$\frac{1}{n} \left[n^2 \left(M_n^{a,b}(f)(v) - f(v) - \frac{[v^{a+1}(1-v)^{b+1} f'(v)]'}{n v^a (1-v)^b} \right) \right] \right\},$$

which replaced in the above Cauchy's formula implies

$$[M_n^{a,b}(f)(z)]^{(p)} - f^{(p)}(z) = \frac{1}{n}\left\{ \frac{p!}{2\pi i}\int_\Gamma \frac{[v^{a+1}(1-v)^{b+1}f'(v)]'/[v^a(1-v)^b]}{(v-z)^{p+1}}dv + \right.$$

$$\frac{1}{n}\cdot\frac{p!}{2\pi i}\int_\Gamma \frac{n^2\left(M_n^{a,b}(f)(v) - f(v) - \frac{[v^{a+1}(1-v)^{b+1}f'(v)]'}{nv^a(1-v)^b}\right)}{(v-z)^{p+1}}dv \left.\right\} =$$

$$\frac{1}{n}\left\{ \left[\frac{(z^{a+1}(1-z)^{b+1}f'(z))'}{z^a(1-z)^b}\right]^{(p)} + \right.$$

$$\frac{1}{n}\cdot\frac{p!}{2\pi i}\int_\Gamma \frac{n^2\left(M_n^{a,b}(f)(v) - f(v) - \frac{[v^{a+1}(1-v)^{b+1}f'(v)]'}{nv^a(1-v)^b}\right)}{(v-z)^{p+1}}dv \left.\right\}.$$

Passing now to $\|\cdot\|_r$, it follows

$$\|[M_n^{a,b}(f)]^{(p)} - f^{(p)}\|_r \geq \frac{1}{n}\left\{ \left\|\left[\frac{(e_1^{a+1}(1-e_1)^{b+1}f')'}{e_1^a(1-e_1)^b}\right]^{(p)}\right\|_r - \right.$$

$$\frac{1}{n}\left\|\frac{p!}{2\pi}\int_\Gamma \frac{n^2\left(M_n^{a,b}(f)(v) - f(v) - \frac{[v^{a+1}(1-v)^{b+1}f'(v)]'}{nv^a(1-v)^b}\right)}{(v-z)^{p+1}}dv\right\|_r \left.\right\},$$

where by using Theorem 1.6.4 we get

$$\left\|\frac{p!}{2\pi}\int_\Gamma \frac{n^2\left(M_n^{a,b}(f)(v) - f(v) - \frac{[v^{a+1}(1-v)^{b+1}f'(v)]'}{nv^a(1-v)^b}\right)}{(v-z)^{p+1}}dv\right\|_r \leq$$

$$\frac{p!}{2\pi}\cdot\frac{2\pi r_1 n^2}{(r_1-r)^{p+1}}\left\|M_n^{a,b}(f) - f - \frac{[e_1^{a+1}(1-e_1)^{b+1}f']'}{ne_1^a(1-e_1)^b}\right\|_{r_1} \leq \frac{M_{r_1,a,b}(f)p!r_1}{(r_1-r)^{p+1}}.$$

But by hypothesis on f we have $\left\|\left[\frac{(e_1^{a+1}(1-e_1)^{b+1}f')'}{e_1^a(1-e_1)^b}\right]^{(p)}\right\|_r > 0.$

Indeed, supposing the contrary it follows that $\frac{[z^{a+1}(1-z)^{b+1}f'(z)]'}{z^a(1-z)^b}$ is a polynomial of degree $\leq p-1$.

But by simple calculation we get

$$\frac{[z^{a+1}(1-z)^{b+1}f'(z)]'}{z^a(1-z)^b} = z(1-z)f''(z) + f'(z)[a+1-z(2+a+b)],$$

therefore it would follow that $z(1-z)f''(z) + f'(z)[a+1-z(2+a+b)] = Q_{p-1}(z)$, where $Q_{p-1}(z) = A_{p-1}z^{p-1} + \ldots + A_0$ is a polynomial of degree $\leq p-1$.

Now, by considering in the above differential equation instead of the complex variable z with $|z| \leq r$ the real variable $x \in [-r,r]$, with $r \geq 1$, by the general theory it easily follows that its general solution is a polynomial of degree $\leq p - 1$ in x (with real coefficients).

Indeed, denoting $Y = y'$ and solving the real homogenous differential equation $x(1 - x)Y'(x) + Y(x)[a + 1 - x(2 + a + b)] = 0$, $x \in [-r,r]$, by simple calculation it easily follows its form of the general solution $Y(x) = \frac{C}{x^{a+1}(1-x)^{b+1}}$. But because $Y(x)$ is differentiable at $x = 0$ and $x = 1$, we necessarily get $C = 0$ and therefore $Y(x) = 0$, so that the general solution, $Y(x)$, of the unhomogenous equation $x(1 - x)Y'(x) + Y(x)[a + 1 - x(2 + a + b)] = Q_{p-1}(x)$ will be the general solution of the homogenous equation plus a particular solution of the above unhomogenous equation, which obviously is a polynomial of degree $\leq p - 2$. In conclusion, $y'(x)$ necessarily is a polynomial of degree $\leq p - 2$, that is, $y(x)$ necessarily is a polynomial of degree $\leq p - 1$; let us denote it by $S_{p-1}(x)$, $x \in [-r,r]$. But then, as $f(z)$ is the analytic extension from $[-r,r]$ to the disk $|z| \leq r$, from the uniqueness theorem on analytic functions, it follows that $f(z) = S_{p-1}(z)$, contradicting the hypothesis on f.

In continuation, reasoning exactly as in the proof of Theorem 1.6.5, we immediately get the desired conclusion. $\qquad\Box$

Remark. For $a = b = 0$ we recapture the results in the case of classical complex Durrmeyer operators studied in the previous section.

1.7 Lorentz Polynomials

In this section we obtain quantitative estimate in the Voronovskaja's theorem and the exact orders in simultaneous approximation by the complex Lorentz polynomials attached to analytic functions in compact disks. Also, we study the approximation properties of their iterates.

These polynomials were introduced in Lorentz [96], p. 43, formula (2), under the name of degenerate Bernstein polynomials, by the formula, attached to any analytic function f in a domain containing the origin,

$$L_n(f)(z) = \sum_{k=0}^{n} \binom{n}{k} \left(\frac{z}{n}\right)^k f^{(k)}(0), n \in \mathbb{N}.$$

In the same book of Lorentz [96], at pages 121–124, some qualitative approximation results are studied.

The first main result of this section is the following.

Theorem 1.7.1 (Gal [52]). *For $R > 1$ and denoting $\mathbb{D}_R = \{z \in \mathbb{C}; |z| < R\}$, suppose that $f : \mathbb{D}_R \to \mathbb{C}$ is analytic in $\mathbb{D}_R$, i.e., $f(z) = \sum_{k=0}^{\infty} c_k z^k$, for all $z \in \mathbb{D}_R$:*

(i) Let $1 \leq r < R$ be arbitrary fixed. For all $|z| \leq r$ and $n \in \mathbb{N}$, we have the upper estimate

$$|L_n(f)(z) - f(z)| \leq \frac{M_r(f)}{n},$$

where $M_r(f) = \frac{1}{2} \sum_{k=2}^{\infty} |c_k| k(k-1) r^k < \infty.$

(ii) For the simultaneous approximation by complex Lorentz polynomials, we have the following: if $1 \leq r < r_1 < R$ are arbitrary fixed, then for all $|z| \leq r$, $p \in \mathbb{N}$ and $n \in \mathbb{N}$, we have

$$|L_n^{(p)}(f)(z) - f^{(p)}(z)| \leq \frac{p! r_1 M_{r_1}(f)}{n(r_1 - r)^{p+1}},$$

where $M_{r_1}(f)$ is given as at the above point (i).

Proof. (i) Denoting $e_j(z) = z^j$, first we easily get that $L_n(e_0)(z) = 1$, $L_n(e_1)(z) = e_1(z)$ and that for all $j, n \in \mathbb{N}$, $2 \leq j \leq n$, we have

$$L_n(e_j)(z) = \binom{n}{j} j! \cdot \frac{z^j}{n^j} = z^j \left(1 - \frac{1}{n}\right) \left(1 - \frac{2}{n}\right) \ldots \left(1 - \frac{j-1}{n}\right).$$

Also, note that for $j \geq n+1$, we easily get $L_n(e_j)(z) = 0$.

Now, since an easy computation shows that

$$L_n(f)(z) = \sum_{j=0}^{\infty} c_j L_n(e_j)(z), \text{ for all } |z| \leq r,$$

we immediately obtain

$$|L_n(f)(z) - f(z)|$$

$$\leq \sum_{j=0}^{n} |c_j| \cdot |L_n(e_j)(z) - e_j(z)| + \sum_{j=n+1}^{\infty} |c_j| \cdot |L_n(e_j)(z) - e_j(z)|$$

$$\leq \sum_{j=2}^{n} |c_j| r^j \left| \left(1 - \frac{1}{n}\right) \left(1 - \frac{2}{n}\right) \ldots \left(1 - \frac{j-1}{n}\right) - 1 \right| + \sum_{j=n+1}^{\infty} |c_j| r^j,$$

for all $|z| \leq r$.

Taking into account the simple inequality $1 - \Pi_{j=1}^{k-1} x_j \leq \sum_{j=1}^{k-1} (1 - x_j)$, valid if $0 \leq x_j \leq 1$, for all $j = 1, \ldots, k-1$, by taking $x_j = 1 - \frac{j}{n}$, we obtain

$$1 - \left(1 - \frac{1}{n}\right) \left(1 - \frac{2}{n}\right) \ldots \left(1 - \frac{k-1}{n}\right) \leq \sum_{j=1}^{k-1} \left[1 - \frac{n-j}{n}\right] = \frac{k(k-1)}{2n}.$$

Since also for $j \geq n+1$ we have $j(j-1)/2 \geq (n+1)/2 \geq 1$, this immediately implies that

$$|L_n(f)(z) - f(z)| \leq \sum_{j=2}^{\infty} |c_j| \cdot r^j \cdot \frac{j(j-1)}{2n},$$

which implies the desired estimate.

(ii) Denoting by γ the circle of radius $r_1 > r$ and center 0, since for any $|z| \leq r$ and $v \in \gamma$, we have $|v - z| \geq r_1 - r$, by Cauchy's formulas it follows that for all $|z| \leq r$ and $n \in \mathbb{N}$, we have

$$|L_n^{(p)}(f)(z) - f^{(p)}(z)| = \frac{p!}{2\pi} \left| \int_\gamma \frac{L_n(f)(v) - f(v)}{(v-z)^{p+1}} dv \right| \leq$$

$$\frac{M_{r_1}(f)}{n} \frac{p!}{2\pi} \cdot \frac{2\pi r_1}{(r_1 - r)^{p+1}} = \frac{M_{r_1}(f)}{n} \cdot \frac{p! r_1}{(r_1 - r)^{p+1}},$$

which proves (ii) and the theorem. □

The following Voronovskaja-type result holds.

Theorem 1.7.2 (Gal [52]). *For $R > 1$, let $f : \mathbb{D}_R \to \mathbb{C}$ be analytic in $\mathbb{D}_R$, that is, $f(z) = \sum_{k=0}^{\infty} c_k z^k$ for all $z \in \mathbb{D}_R$, and let $1 \leq r < R$ be arbitrary fixed. For all $|z| \leq r$ we have*

$$\left| L_n(f)(z) - f(z) + \frac{z^2}{2n} f''(z) \right| \leq \frac{1}{2n^2} \sum_{k=2}^{\infty} |c_k| r^k (k-1)^2 (k-2)^2, \text{ for all } n \in \mathbb{N},$$

where $\sum_{k=2}^{\infty} |c_k| r^k (k-1)^2 (k-2)^2 < \infty$.

Proof. We have

$$\left| L_n(f)(z) - f(z) + \frac{z^2}{2n} f''(z) \right| = \left| \sum_{k=0}^{\infty} c_k \left[L_n(e_k)(z) - e_k(z) + \frac{k(k-1)}{2n} e_k(z) \right] \right|$$

$$\leq \left| \sum_{k=2}^{n} c_k z^k \left[\frac{(n-1)(n-2)\dots(n-(k-1))}{n^{k-1}} - 1 + \frac{k(k-1)}{2n} \right] \right|$$

$$+ \left| \sum_{k=n+1}^{\infty} c_k z^k \left[-1 + \frac{k(k-1)}{2n} \right] \right|$$

$$\leq \sum_{k=2}^{\infty} |c_k| r^k \left| \frac{(n-1)(n-2)\dots(n-(k-1))}{n^{k-1}} - 1 + \frac{k(k-1)}{2n} \right|,$$

for all $|z| \leq r$ and $n \in \mathbb{N}$.

In what follows, firstly we will prove by mathematical induction with respect to k that

$$0 \leq E_{n,k} \leq \frac{(k-1)^2(k-2)^2}{2n^2}, \tag{1.7.1}$$

for all $k \geq 2$ (here $n \in \mathbb{N}$ is arbitrary fixed), where

$$E_{n,k} = \frac{(n-1)(n-2)\ldots(n-(k-1))}{n^{k-1}} - 1 + \frac{k(k-1)}{2n}. \tag{1.7.2}$$

Indeed, for $k = 2$ it is trivial. Suppose that it is valid for arbitrary k. We are going to prove that it remains valid for $k+1$ too, that is,

$$0 \leq \frac{(n-1)(n-2)\ldots(n-k)}{n^k} - 1 + \frac{k(k+1)}{2n} \leq \frac{k^2(k-1)^2}{2n^2}. \tag{1.7.3}$$

For this purpose, we take into account that

$$E_{n,k+1} = \frac{(n-1)(n-2)\ldots(n-k)}{n^k} - 1 + \frac{k(k+1)}{2n}$$

$$= \frac{(n-1)(n-2)\ldots(n-(k-1))}{n^{k-1}}\left(1 - \frac{k}{n}\right) - 1 + \frac{k(k-1)}{2n} + \frac{k(k+1)}{2n} - \frac{k(k-1)}{2n}$$

$$= E_{n,k} + \frac{k}{n}\left(1 - \frac{(n-1)(n-2)\ldots(n-(k-1))}{n^{k-1}}\right).$$

By (1.7.2) it is immediate that $E_{n,k+1} \geq 0$. Also, by the relationship (1.7.1) and taking into account the simple inequality used at the end of the proof of Theorem 1.7.1, (i), we get

$$E_{n,k+1} \leq \frac{(k-1)^2(k-2)^2}{2n^2} + \frac{k}{n}\left(1 - \frac{(n-1)(n-2)\ldots(n-(k-1))}{n^{k-1}}\right)$$

$$\leq \frac{(k-1)^2(k-2)^2}{2n^2} + \frac{k}{n} \cdot \frac{k(k-1)}{2n} = \frac{1}{2n^2}[(k-1)^2(k-2)^2 + k^2(k-1)].$$

Looking at (1.7.3), in fact it remains to prove that

$$(k-1)^2(k-2)^2 + k^2(k-1) \leq k^2(k-1)^2,$$

which after simple calculation is equivalent to the inequality $0 \leq 3k^2 - 8k + 4$, which is obviously valid for all $k \geq 2$.

In conclusion, (1.7.2) is valid, which implies that (1.7.1) is valid and it is immediate that $E_{n,k} \leq \frac{(k+1)^2(k+2)^2}{2n^2}$. $\qquad\square$

To obtain the exact approximation order, first we need the following.

Theorem 1.7.3 (Gal [52]). *Let $R > 1$, $f : \mathbb{D}_R \to \mathbb{C}$ be analytic in $\mathbb{D}_R$, that is, $f(z) = \sum_{k=0}^{\infty} c_k z^k$ for all $z \in \mathbb{D}_R$, and $1 \leq r < R$ be arbitrary fixed. If f is not a polynomial of degree ≤ 1, then for all $n \in \mathbb{N}$ and $|z| \leq r$, we have*

$$\|L_n(f) - f\|_r \geq \frac{C_r(f)}{n},$$

where the constant $C_r(f)$ depends only on f. Here $\|f\|_r$ denotes $\max_{|z| \le r}\{|f(z)|\}$.

Proof. For all $|z| \le r$ and $n \in \mathbb{N}$ we have

$$L_n(f)(z) - f(z) =$$

$$\frac{1}{n}\left\{-\frac{z^2}{2}f''(z) + \frac{1}{n}\left[n^2\left(L_n(f)(z) - f(z) + \frac{z^2}{2n}f''(z)\right)\right]\right\}.$$

In what follows we will apply to this identity the following obvious property:

$$\|F + G\|_r \ge |\,\|F\|_r - \|G\|_r\,| \ge \|F\|_r - \|G\|_r.$$

It follows

$$\|L_n(f) - f\|_r \ge \frac{1}{n}\left\{\left\|\frac{e_1^2}{2}f''\right\|_r - \frac{1}{n}\left[n^2\left\|L_n(f) - f + \frac{e_1^2}{2n}f''\right\|_r\right]\right\}.$$

Since by hypothesis f is not a polynomial of degree ≤ 1 in $\mathbb{D}_R$, we get $\left\|\frac{e_1^2}{2}f''\right\|_r > 0$.

Indeed, supposing the contrary it follows that $\frac{z^2}{2}f''(z) = 0$ for all $z \in \overline{\mathbb{D}}_r = \{z \in \mathbb{C}; |z| \le r\}$, which implies $f''(z) = 0$ for all $z \in \overline{\mathbb{D}}_r \setminus \{0\}$. Since f is supposed to be analytic, from the identity theorem of analytic (holomorphic) functions, this necessarily implies that $f''(z) = 0$, for all $z \in \mathbb{D}_R$, i.e., that f is a polynomial of degree ≤ 1, which is a contradiction.

But by Theorem 1.7.2 we have

$$n^2\left\|L_n(f) - f + \frac{e_1^2}{2n}f''\right\|_r \le \frac{1}{2}\cdot\sum_{k=2}^{\infty}|c_k|\cdot r^k(k-1)^2(k-2)^2.$$

Therefore, there exists an index n_0 depending only on f and r, such that for all $n > n_0$, we have

$$\left\|\frac{e_1^2}{2}f''\right\|_r - \frac{1}{n}\left[n^2\left\|L_n(f) - f + \frac{e_1^2}{2n}f''\right\|_r\right] \ge \frac{1}{2}\left\|\frac{e_1^2}{2}f''\right\|_r,$$

which immediately implies that

$$\|L_n(f) - f\|_r \ge \frac{1}{n}\cdot\frac{1}{2}\left\|\frac{e_1^2}{2}f''\right\|_r, \forall n > n_0.$$

For $n \in \{1,\ldots,n_0\}$ we obviously have $\|L_n(f) - f\|_r \ge \frac{M_{r,n}(f)}{n}$ with $M_{r,n}(f) = n\cdot\|L_n(f) - f\|_r > 0$ (if $\|L_n(f) - f\|_r$ would be equal to 0, this would imply that f is a linear function, a contradiction).

Therefore, finally we get $\|L_n(f) - f\|_r \ge \frac{C_r(f)}{n}$ for all $n \in \mathbb{N}$, where

$$C_r(f) = \min\left\{M_{r,1}(f), \ldots, M_{r,n_0}(f), \frac{1}{2}\left\|\frac{e_1^2}{2}f''\right\|_r\right\},$$

which completes the proof.   □

Combining now Theorem 1.7.3 with Theorem 1.7.1, (i), we immediately get the following exact estimate.

Corollary 1.7.4 (Gal [52]). *Let $R > 1$, $f : \mathbb{D}_R \to \mathbb{C}$ be analytic in $\mathbb{D}_R$, that is, $f(z) = \sum_{k=0}^{\infty} c_k z^k$ for all $z \in \mathbb{D}_R$, and $1 \le r < R$ be arbitrary fixed. If f is not a polynomial of degree ≤ 1, then for all $n \in \mathbb{N}$, we have*

$$\|L_n(f) - f\|_r \sim \frac{1}{n},$$

where the constants in the equivalence depend on f and r but are independent of n.

Concerning the simultaneous approximation, we present the following.

Theorem 1.7.5 (Gal [52]). *Let $R > 1$, $f : \mathbb{D}_R \to \mathbb{C}$ be analytic in $\mathbb{D}_R$, that is, $f(z) = \sum_{k=0}^{\infty} c_k z^k$ for all $z \in \mathbb{D}_R$, and $1 \le r < r_1 < R$ be arbitrary fixed. Also, let $p \in \mathbb{N}$. If f is not a polynomial of degree $\le \max\{1, p-1\}$, then for all $n \in \mathbb{N}$, we have*

$$\|L_n^{(p)}(f) - f^{(p)}\|_r \sim \frac{1}{n},$$

where the constants in the equivalence depend on f, r, r_1, and p but are independent of n.

Proof. Since by Theorem 1.7.1, (ii) and by the Remark after the proof of Theorem 1.7.1, we have the upper estimate for $\|L_n^{(p)}(f) - f^{(p)}\|_r$, it remains to prove the lower estimate for $\|L_n^{(p)}(f) - f^{(p)}\|_r$. For this purpose, denoting by Γ the circle of radius r_1 and center 0, we have the inequality $|v - z| \ge r_1 - r$ valid for all $|z| \le r$ and $v \in \Gamma$. Cauchy's formula is expressed by

$$L_n^{(p)}(f)(z) - f^{(p)}(z) = \frac{p!}{2\pi i}\int_\Gamma \frac{L_n(f)(v) - f(v)}{(v - z)^{p+1}}\,dv.$$

Now, as in the proof of Theorem 1.7.1, (ii), for all $v \in \Gamma$ and $n \in \mathbb{N}$, we have

$$L_n(f)(v) - f(v) =$$

$$\frac{1}{n}\left\{-\frac{v^2}{2}f''(v) + \frac{1}{n}\left[n^2\left(L_n(f)(v) - f(v) + \frac{v^2}{2n}f''(v)\right)\right]\right\},$$

which replaced in the above Cauchy's formula implies

$$L_n^{(p)}(f)(z) - f^{(p)}(z) = \frac{1}{n}\left\{\frac{p!}{2\pi i}\int_\Gamma -\frac{v^2 f''(v)}{2(v-z)^{p+1}}dv+\right.$$

$$\frac{1}{n}\cdot\frac{p!}{2\pi i}\int_\Gamma \frac{n^2\left(L_n(f)(v) - f(v) + \frac{v^2}{2n}f''(v)\right)}{(v-z)^{p+1}}dv\left.\right\} =$$

$$\frac{1}{n}\left\{\left[-\frac{z^2}{2}f''(z)\right]^{(p)} + \frac{1}{n}\cdot\frac{p!}{2\pi i}\int_\Gamma \frac{n^2\left(L_n(f)(v) - f(v) + \frac{v^2}{2n}f''(v)\right)}{(v-z)^{p+1}}dv\right\}.$$

Passing now to $\|\cdot\|_r$, for all $n \in \mathbb{N}$ it follows

$$\|L_n^{(p)}(f) - f^{(p)}\|_r \geq$$

$$\frac{1}{n}\left\{\left\|\left[-\frac{e_1^2}{2}f''\right]^{(p)}\right\|_r - \frac{1}{n}\left\|\frac{p!}{2\pi}\int_\Gamma \frac{n^2\left(L_n(f)(v) - f(v) + \frac{v^2}{2n}f''(v)\right)}{(v-z)^{p+1}}dv\right\|_r\right\},$$

where by using Theorem 1.7.2, for all $n \in \mathbb{N}$ we get

$$\left\|\frac{p!}{2\pi}\int_\Gamma \frac{n^2\left(L_n(f)(v) - f(v) + \frac{v^2}{2n}f''(v)\right)}{(v-z)^{p+1}}dv\right\|_r$$

$$\leq \frac{p!}{2\pi}\cdot\frac{2\pi r_1 n^2}{(r_1 - r)^{p+1}}\left\|L_n(f) - f + \frac{e_1^2}{2n}f''\right\|_{r_1}$$

$$\leq \frac{1}{2}\sum_{k=2}^\infty r_1^k (k-1)^2 (k-2)^2 \cdot \frac{p! r_1}{(r_1 - r)^{p+1}}.$$

But by hypothesis on f, we have $\left\|-\left[\frac{e_1^2}{2}f''\right]^{(p)}\right\|_r > 0$. Indeed, supposing the contrary it follows that $-\frac{z^2}{2}f''(z)$ is a polynomial of degree $\leq p-1$.

Now, if $p = 1$ and $p = 2$, then the analyticity of f obviously implies that f necessarily is a polynomial of degree $\leq 1 = \max\{1, p-1\}$, which contradicts the hypothesis. If $p > 2$, then the analyticity of f obviously implies that f necessarily is a polynomial of degree $\leq p - 1 = \max\{1, p-1\}$, which again contradicts the hypothesis.

In continuation, reasoning exactly as in the proof of Theorem 1.7.3, we immediately get the desired conclusion. $\square$

In what follows, we consider the approximation by iterates. For f analytic in $\mathbb{D}_R$ that is of the form $f(z) = \sum_{k=0}^\infty c_k z^k$, for all $z \in \mathbb{D}_R$, let us define the iterates of complex Lorentz polynomial $L_n(f)(z)$, by $L_n^{(1)}(f)(z) = L_n(f)(z)$ and $L_n^{(m)}(f)(z) = L_n[L_n^{(m-1)}(f)](z)$, for any $m \in \mathbb{N}$, $m \geq 2$.

Since we have $L_n(f)(z) = \sum_{k=0}^{\infty} c_k L_n(e_k)(z)$, by recurrence for all $m \geq 1$, we easily get that

$$L_n^{(m)}(f)(z) = \sum_{k=0}^{\infty} c_k L_n^{(m)}(e_k)(z),$$

where $L_n^{(m)}(e_k)(z) = 1$ if $k = 0$, $L_n^{(m)}(e_k)(z) = z$ if $k = 1$, $L_n^{(m)}(e_k)(z) = 0$ if $k \geq n+1$ and

$$L_n^{(m)}(e_k)(z) = \left(1 - \frac{1}{n}\right)^m \left(1 - \frac{2}{n}\right)^m \cdots \left(1 - \frac{k-1}{n}\right)^m z^k, \text{ for } 2 \leq k \leq n.$$

The main result is the following.

Theorem 1.7.6 (Gal [52]). *Let f be analytic in $\mathbb{D}_R$ with $R > 1$, that is,* $f(z) = \sum_{k=0}^{\infty} c_k z^k$, *for all $z \in \mathbb{D}_R$. Let $1 \leq r < R$.*
We have

$$\|L_n^{(m)}(f) - f\|_r \leq \frac{m}{n} \sum_{k=2}^{\infty} |c_k| \frac{k(k-1)}{2} r^k,$$

and therefore if $\lim_{n \to \infty} \frac{m}{n} = 0$, *then*

$$\lim_{n \to \infty} \|L_n^{(m)}(f) - f\|_r = 0.$$

Proof. For all $|z| \leq r$, we easily obtain

$$|f(z) - L_n^{(m)}(f)(z)| \leq \sum_{k=2}^{n} |c_k| r^k \left[1 - \left(1 - \frac{1}{n}\right)^m \left(1 - \frac{2}{n}\right)^m \cdots \left(1 - \frac{k-1}{n}\right)^m\right]$$

$$+ \sum_{k=n+1}^{\infty} |c_k| r^k.$$

Denoting $A_k = \left(1 - \frac{1}{n}\right)\left(1 - \frac{2}{n}\right) \cdots \left(1 - \frac{k-1}{n}\right)$, we get $1 - A_k^m = (1 - A_k)(1 + A + A^2 + \ldots + A^{m-1}) \leq m(1 - A_k)$, and therefore since $1 - A_k \leq \frac{k(k-1)}{2n}$, for all $|z| \leq r$, we obtain

$$|f(z) - L_n^{(m)}(f)(z)| \leq m \sum_{k=2}^{n} |c_k| r^k [1 - A_k] + m \sum_{k=n+1}^{\infty} |c_k| \cdot r^k$$

$$\leq \frac{m}{n} \sum_{k=2}^{n} |c_k| r^k \frac{k(k-1)}{2} + m \sum_{k=n+1}^{\infty} |c_k| \cdot r^k \cdot \frac{k(k-1)}{2n}$$

$$\leq \frac{m}{n} \sum_{k=2}^{\infty} |c_k| r^k \cdot \frac{k(k-1)}{2},$$

which immediately proves the theorem. $\qquad\square$

Finally, we present the preservation of some geometric properties through the Lorentz polynomials. For this purpose, we introduce the following two classes of functions:

$S_1 = \{f : \mathbb{D}_1 \to \mathbb{C}; \ f(z) = z + c_2 z^2 + \ldots, \text{ analytic in } \mathbb{D}_1, \text{ satisfying } \sum_{k=2}^{\infty} k|c_k| \leq 1\}$,

and

$S_2 = \{f : \mathbb{D}_1 \to \mathbb{C}; \ f(z) = c_1 z + c_2 z^2 + \ldots, \text{ analytic in } \mathbb{D}_1, \text{ satisfying } |c_1| \geq \sum_{k=2}^{\infty} |c_k|\}$.

According to, e.g., Mocanu, Bulboacă, and Sălăgean [110], p. 97, Exercise 4.9.1, if $f \in S_1$, then $|\frac{zf'(z)}{f(z)} - 1| < 1, z \in \mathbb{D}_1$ and therefore f is starlike (univalent) on $\mathbb{D}_1$.

Also, according to Alexander [5], p. 22, if $f \in S_2$, then f is starlike (and univalent) in $\mathbb{D}_1$. Therefore S_1 and S_2 are subsets of the class of univalent starlike functions on $\mathbb{D}_1$.

We have the following:

Theorem 1.7.7. *If $f \in S_1$, then $L_n(f) \in S_1$ for all $n \in \mathbb{N}$, and if $f \in S_2$, then $L_n(f) \in S_2$ for all $n \in \mathbb{N}$.*

Proof. Let first $f \in S_1$, that is, $f(z) = z + c_2 z^2 + \ldots,$. By the proof of Theorem 1.7.1, we can write

$$L_n(f)(z) = \sum_{j=0}^{\infty} A_{j,n} z^j,$$

where $A_{j,n} = c_j(1 - 1/n) \ldots (1 - (j-1)/n)$ if $0 \leq j \leq n$ and $A_{j,n} = 0$ for $j \geq n+1$, $(1 - 1/n) \ldots (1 - (j-1)/n) = 1$ for $j = 0$ and $j = 1$ and c_j are the coefficients of the development in series of f, that is, $c_0 = 0$, $c_1 = 1$. It follows $A_{0,n} = 0$, $A_{1,n} = 1$, and we get

$$\sum_{k=2}^{\infty} k|A_{k,n}| = \sum_{k=2}^{n} k|c_k|(1 - 1/n) \ldots (1 - (k-1)/n) \leq \sum_{k=2}^{\infty} k|c_k| \leq 1,$$

which shows that $L_n(f) \in S_1$ for all $n \in \mathbb{N}$.

Now, if $f \in S_2$, that is, $f(z) = c_1 z + c_2 z^2 + \ldots$, we get $A_{0,n} = 0$, $A_{1,n} = c_1$ and

$$|A_{1,n}| = |c_1| \geq \sum_{k=2}^{\infty} |c_k| \geq \sum_{k=2}^{\infty} |A_{k,n}|,$$

which shows that $L_n(f) \in S_2$ for all $n \in \mathbb{N}$. $\square$

1.8 q-Lorentz Polynomials, $q > 1$

In this section, for $q > 1$, we obtain quantitative estimate in the Voronovskaja's theorem and the exact orders in simultaneous approximation by the complex q-Lorentz polynomial of degree $n \in \mathbb{N}$, attached to analytic functions in compact disks of the complex plane. The error of approximation attained is $\frac{1}{[n]_q}$, which, by the inequalities $(q-1)\frac{1}{q^n} \leq \frac{1}{[n]_q} \leq q\frac{1}{q^n}$, implies the geometric progression order of approximation q^{-n}, essentially improving the approximation order $1/n$ for the case $q = 1$, obtained in the very recent paper of Gal [52] (see also the previous Sect. 1.7). Moreover, some approximation properties of the iterates of these complex q-polynomials are studied.

We define the complex q-Lorentz polynomials, $q \geq 1$, by

$$L_{n,q}(f)(z) = \sum_{k=0}^{n} q^{k(k-1)/2} \binom{n}{k}_q \left(\frac{z}{[n]_q}\right)^k D_q^{(k)}(f)(0), n \in \mathbb{N}, z \in \mathbb{C}.$$

Recall here $[n]_q = \frac{q^n-1}{q-1}$ if $q \neq 1$, $[n]_q = n$ if $q = 1$, $[n]_q! = [1]_q[2]_q \cdots \cdots$ $[n]_q$, $\binom{n}{k}_q = \frac{[n]_q!}{[k]_q![n-k]_q!}$, $D_q^k(f)(z) = D_q[D_q^{k-1}(f)](z)$, $D_q^0(f)(z) = f(z)$, $D_q^1(f)(z) = D_q(f)(z) = \frac{f(qz)-f(z)}{(q-1)z}$.

Note that because $D_q(e_k)(z) = [k]_q z^{k-1}$, where $e_k(z) = z^k$, if f is analytic in a disk $\mathbb{D}_R = \{z \in \mathbb{C}; |z| < R\}$, that is, we have $f(z) = \sum_{k=0}^{\infty} c_k z^k$ for all $z \in \mathbb{D}_R$, then $D_q(f)(z) = \sum_{k=1}^{\infty} c_k[k]_q z^{k-1}$, $D_q^2(f)(z) = \sum_{k=2}^{\infty} c_k[k]_q[k-1]_q z^{k-2}$, and so on. This immediately implies that $D_q^k(f)(0) = c_k[k]_q!$, for all $k = 0, 1, \ldots,$. Also, since $[n]_1 = n$ and $D_1(f)(z) = f'(z)$, it is immediate that $L_{n,1}(f)(z)$ becomes the complex original Lorentz polynomials $L_n(f)(z)$ defined and studied in Sect. 1.7.

Firstly we obtain an upper approximation estimate.

Theorem 1.8.1 (Gal [58]). *Let $R > q > 1$. Denoting $\mathbb{D}_R = \{z \in \mathbb{C}; |z| < R\}$, suppose that $f : \mathbb{D}_R \to \mathbb{C}$ is analytic in $\mathbb{D}_R$, i.e., $f(z) = \sum_{k=0}^{\infty} c_k z^k$, for all $z \in \mathbb{D}_R$:*

(i) Let $1 \leq r < \frac{r_1}{q} < \frac{R}{q}$ be arbitrary fixed. For all $|z| \leq r$ and $n \in \mathbb{N}$, we have the upper estimate

$$|L_{n,q}(f)(z) - f(z)| \leq \frac{M_{r_1,q}(f)}{[n]_q},$$

where $M_{r_1,q}(f) = \frac{q+1}{(q-1)^2} \cdot \sum_{k=0}^{\infty} |c_k|(k+1)r_1^k < \infty$.

(ii) Let $1 \leq r < r^* < \frac{r_1}{q} < \frac{R}{q}$ be arbitrary fixed. For the simultaneous approximation by complex Lorentz polynomials, for all $|z| \leq r$, $p \in \mathbb{N}$ and $n \in \mathbb{N}$, we have

$$|L_{n,q}^{(p)}(f)(z) - f^{(p)}(z)| \leq \frac{p! r^* M_{r_1,q}(f)}{[n]_q (r^* - r)^{p+1}},$$

where $M_{r_1,q}(f)$ is given as at the above point (i).

Proof. (i) Denoting $e_j(z) = z^j$, firstly we easily get that $L_{n,q}(e_0)(z) = 1$, $L_{n,q}(e_1)(z) = e_1(z)$. Then, since for all $j, n \in \mathbb{N}$, $2 \leq j \leq n$, we have

$$L_{n,q}(e_j)(z) = q^{j(j-1)/2} \binom{n}{j}_q [j]_q! \cdot \frac{z^j}{[n]_q^j},$$

taking into account the relationship (7), p. 236 in Ostrovska [117], we get

$$L_{n,q}(e_j)(z) = z^j \left(1 - \frac{[1]_q}{[n]_q}\right) \left(1 - \frac{[2]_q}{[n]_q}\right) \cdots \left(1 - \frac{[j-1]_q}{[n]_q}\right).$$

Also, note that for $j \geq n+1$, we easily get $L_{n,q}(e_j)(z) = 0$.
Now, since an easy computation shows that

$$L_{n,q}(f)(z) = \sum_{j=0}^{\infty} c_j L_{n,q}(e_j)(z), \text{ for all } |z| \leq r,$$

we immediately obtain

$$|L_{n,q}(f)(z) - f(z)|$$

$$\leq \sum_{j=0}^{n} |c_j| \cdot |L_{n,q}(e_j)(z) - e_j(z)| + \sum_{j=n+1}^{\infty} |c_j| \cdot |L_{n,q}(e_j)(z) - e_j(z)|$$

$$\leq \sum_{j=2}^{n} |c_j| r^j \left| \left(1 - \frac{[1]_q}{[n]_q}\right) \left(1 - \frac{[2]_q}{[n]_q}\right) \cdots \left(1 - \frac{[j-1]_q}{[n]_q}\right) - 1 \right| + \sum_{j=n+1}^{\infty} |c_j| r^j,$$

for all $|z| \leq r$.
Taking into account the inequality proved in Ostrovska [117], p. 247

$$1 - \left(1 - \frac{[1]_q}{[n]_q}\right) \left(1 - \frac{[2]_q}{n}\right) \cdots \left(1 - \frac{[j-1]_q}{n}\right) \leq \frac{(j-1)[j-1]_q}{[n]_q},$$

we obtain

$$\sum_{j=2}^{n} |c_j| r^j \left| \left(1 - \frac{[1]_q}{[n]_q}\right) \left(1 - \frac{[2]_q}{[n]_q}\right) \cdots \left(1 - \frac{[j-1]_q}{[n]_q}\right) - 1 \right|$$

$$\leq \frac{1}{[n]_q} \sum_{j=2}^{\infty} |c_j| (j-1)[j-1]_q r^j \leq \frac{1}{[n]_q} \sum_{j=2}^{\infty} |c_j| \cdot \frac{j q^j}{q-1} \cdot r^j$$

$$\leq \frac{1}{[n]_q} \cdot \frac{1}{q-1} \sum_{j=2}^{\infty} |c_j|(j+1)(rq)^j \leq \frac{1}{[n]_q} \cdot \frac{1}{q-1} \sum_{j=2}^{\infty} |c_j|(j+1)r_1^j,$$

where by hypothesis on f we have $\sum_{j=0}^{\infty} |c_j|(j+1)r_1^j < \infty$.

On the other hand, the analyticity of f implies $c_j = \frac{f^{(k)}(0)}{j!}$, and by Cauchy's estimates of the coefficients c_j in the disk $|z| \leq r_1$, we have $|c_j| \leq \frac{K_{r_1}}{r_1^j}$, for all $j \geq 0$, where

$$K_{r_1} = \max\{|f(z)|; |z| \leq r_1\} \leq \sum_{j=0}^{\infty} |c_j|r_1^j \leq \sum_{j=0}^{\infty} |c_j|(j+1)r_1^j := R_{r_1}(f) < \infty.$$

Therefore we get

$$\sum_{j=n+1}^{\infty} |c_j|r^j \leq R_{r_1}(f)\left[\frac{r}{r_1}\right]^{n+1} \sum_{j=0}^{\infty} \left(\frac{r}{r_1}\right)^j = R_{r_1}(f)\left[\frac{r}{r_1}\right]^{n+1} \cdot \frac{r_1}{r_1-r}$$

$$= R_{r_1}(f) \cdot \frac{r}{r_1-r} \cdot \left[\frac{r}{r_1}\right]^n \leq \frac{R_{r_1}(f)}{q-1} \cdot \left[\frac{r}{r_1}\right]^n \leq \frac{R_{r_1}(f)}{q-1} \cdot \frac{1}{q^n} \leq \frac{2R_{r_1}(f)}{(q-1)^2} \cdot \frac{1}{[n]_q}.$$

Collecting the estimates, finally we obtain

$$|L_{n,q}(f)(z)-f(z)| \leq \frac{1}{[n]_q} \cdot \frac{R_{r_1}(f)}{q-1}\left(1 + \frac{2}{q-1}\right) = \frac{1}{[n]_q} \cdot \frac{q+1}{(q-1)^2} \cdot \sum_{j=0}^{\infty} |c_j|(j+1)r_1^j,$$

for all $n \in \mathbb{N}$ and $|z| \leq r$.

(ii) Denoting by γ the circle of radius $r^* > r$ and center 0, since for any $|z| \leq r$ and $v \in \gamma$, we have $|v - z| \geq r^* - r$, by Cauchy's formulas it follows that for all $|z| \leq r$ and $n \in \mathbb{N}$, we have

$$|L_{n,q}^{(p)}(f)(z) - f^{(p)}(z)| = \frac{p!}{2\pi}\left| \int_{\gamma} \frac{L_{n,q}(f)(v) - f(v)}{(v - z)^{p+1}} dv \right| \leq$$

$$\frac{M_{r_1,q}(f)}{[n]_q} \frac{p!}{2\pi} \cdot \frac{2\pi r^*}{(r^* - r)^{p+1}} = \frac{M_{r_1,q}(f)}{[n]_q} \cdot \frac{p!r^*}{(r^* - r)^{p+1}},$$

which proves (ii) and the theorem. $\qquad \square$

The following quantitative Voronovskaja-type result holds.

Theorem 1.8.2 (Gal [58]). *For $R > q^4 > 1$ let $f : \mathbb{D}_R \to \mathbb{C}$ be analytic in $\mathbb{D}_R$, that is, $f(z) = \sum_{k=0}^{\infty} c_k z^k$ for all $z \in \mathbb{D}_R$, and let $1 \leq r < \frac{r_1}{q^3} < \frac{R}{q^4}$ be arbitrary fixed. For all $n \in \mathbb{N}, |z| \leq r$, we have*

$$\left| L_{n,q}(f)(z) - f(z) + \frac{P_q(f)(z)}{[n]_q} \right| \leq \frac{Q_{r_1,q}(f)}{[n]_q^2},$$

where

$$P_q(f)(z) = \sum_{k=2}^{\infty} c_k \frac{[k]_q - k}{q - 1} z^k = \sum_{k=2}^{\infty} c_k([1]_q + \ldots + [k-1]_q) z^k,$$

and $Q_{r_1,q}(f) = \frac{q^2 - 2q + 2}{(q-1)^3} \cdot \sum_{k=0}^{\infty} |c_k|(k+1)(k+2)^2 (r_1 q)^k < \infty.$

Proof. We have

$$\left| L_{n,q}(f)(z) - f(z) + \frac{P_q(f)(z)}{[n]_q} \right|$$

$$= \left| \sum_{k=0}^{\infty} c_k \left[L_{n,q}(e_k)(z) - e_k(z) + \frac{[k]_q - k}{(q-1)[n]_q} e_k(z) \right] \right|$$

$$\leq \left| \sum_{k=0}^{n} c_k \left[L_{n,q}(e_k)(z) - e_k(z) + \frac{[k]_q - k}{(q-1)[n]_q} e_k(z) \right] \right|$$

$$+ \left| \sum_{k=n+1}^{\infty} c_k z^k \left(\frac{[k] - k}{(q-1)[n]_q} - 1 \right) \right|$$

$$\leq \left| \sum_{k=0}^{n} c_k \left[L_{n,q}(e_k)(z) - e_k(z) + \frac{[k]_q - k}{(q-1)[n]_q} e_k(z) \right] \right|$$

$$+ \sum_{k=n+1}^{\infty} |c_k| r^k \left(\frac{[k]_q - k}{(q-1)[n]_q} - 1 \right),$$

for all $|z| \leq r$ and $n \in \mathbb{N}$.

In what follows, firstly we will prove by mathematical induction with respect to k that

$$0 \leq E_{n,k,q}(z) \leq \frac{r_1^2}{[n]_q^2} (k-1)(k-2)^2 [k-2]_q, \tag{1.8.1}$$

for all $2 \leq k \leq n$ (here $n \in \mathbb{N}$ is arbitrary fixed) and $|z| \leq r$, where

$$E_{n,k,q}(z) = L_{n,q}(e_k)(z) - e_k(z) + \frac{[k]_q - k}{(q-1)[n]_q} e_k(z)$$

$$= L_{n,q}(e_k)(z) - e_k(z) + \frac{1}{[n]_q} ([1]_q + \ldots + [k-1]_q) e_k(z).$$

Note that the relationship

$$\frac{[k]_q - k}{(q-1)[n]_q} e_k(z) = \frac{1}{[n]_q} ([1]_q + \ldots + [k-1]_q) e_k(z), \ k \geq 2,$$

easily follows by mathematical induction.

On the other hand, by the formula for $L_{n,q}(e_k)$ in the proof of Theorem 1.8.1, (i), simple calculation leads to $E_{n,2,q}(z) = 0$, for all $n \in \mathbb{N}$ and to the recurrence formulas

$$L_{n,q}(e_{j+1})(z) = -\frac{z^2}{[n]_q} D_q\left[L_{n,q}(e_j)\right](z) + z L_{n,q}(e_j)(z), \, j \geq 1, n \in \mathbb{N}, |z| \leq r,$$

and

$$E_{n,k,q}(z) = -\frac{z^2}{[n]_q} D_q(L_{n,q}(e_{k-1})(z) - z^{k-1}) + z E_{n,k-1,q}(z), \, n \geq k \geq 3, |z| \leq r.$$

Passing to absolute value above with $|z| \leq r$ and $3 \leq k \leq n$ and applying the mean value theorem in complex analysis, with the general notation $\|f\|_r = \max\{|f(z)|; |z| \leq r\}$, one obtains

$$|E_{n,k,q}(z)| \leq \frac{r^2}{[n]_q} \|(L_{n,q}(e_{k-1})(z) - z^{k-1})'\|_{qr} + r \cdot |E_{n,k-1,q}(z)|$$

$$\leq r \cdot |E_{n,k-1,q}(z)| + \frac{r^2}{[n]_q} \cdot \frac{k-1}{qr} \|L_{n,q}(e_{k-1})(z) - z^{k-1}\|_{qr}$$

$$\leq r \cdot |E_{n,k-1,q}(z)| + \frac{r^2}{[n]_q} \cdot \frac{k-1}{qr} \cdot (qr)^{k-1} \cdot \frac{(k-2)[k-2]_q}{[n]_q},$$

where above we used the estimate which easily follows from the proof of Theorem 1.8.1, (i):

$$|L_{n,q}(e_k)(z) - z^k| \leq r^k \frac{(k-1)[k-1]_q}{[n]_q}, \, |z| \leq r, k \geq 2.$$

Therefore, for all $|z| \leq r$, $3 \leq k \leq n$, we got

$$|E_{n,k,q}(z)| \leq r \cdot |E_{n,k-1,q}(z)| + \frac{r^2}{[n]_q} \cdot \frac{k-1}{qr} \cdot (qr)^{k-1} \cdot \frac{(k-2)[k-2]_q}{[n]_q}$$

$$\leq r_1 \cdot |E_{n,k-1,q}(z)| + \frac{r_1^2}{[n]_q} \cdot (k-1) r_1^{k-2} \cdot \frac{(k-2)[k-2]_q}{[n]_q}$$

$$= r_1 \cdot |E_{n,k-1,q}(z)| + \frac{(k-1)(k-2)[k-2]_q}{[n]_q^2} \cdot r_1^k.$$

Taking $k = 3, 4, \ldots$, step by step, we easily obtain the estimate

$$|E_{n,k,q}(z)| \leq \frac{r_1^k}{[n]_q^2} \left(\sum_{j=3}^{k} (j-1)(j-2)[j-2]_q\right) \leq \frac{r_1^k}{[n]_q^2} (k-1)(k-2)^2 [k-2]_q$$

$$\leq \frac{(r_1 q)^k}{(q-1)[n]_q^2} (k-1)(k-2)^2,$$

for all $|z| \leq r$ and $3 \leq k$ because $[k-2]_q \leq \frac{q^k}{q-1}$.

In conclusion, (1.8.1) is valid, which implies

$$\left| \sum_{k=0}^{n} c_k \left[L_{n,q}(e_k)(z) - e_k(z) + \frac{[k]_q - k}{(q-1)[n]_q} e_k(z) \right] \right| \leq \sum_{k=0}^{n} |c_k| \cdot |E_{n,k,q}(z)|$$

$$\leq \frac{1}{(q-1)[n]_q^2} \sum_{k=3}^{n} |c_k|(k-1)(k-2)^2(r_1 q)^k$$

$$\leq \frac{1}{(q-1)[n]_q^2} \sum_{k=0}^{\infty} |c_k|(k+1)(k+2)^2(r_1 q)^k.$$

On the other hand, since $\frac{[k]_q - k}{(q-1)[n]_q} - 1 \geq 0$ for all $k \geq n+1$, reasoning exactly as in the proof of Theorem 1.8.1, (i), and keeping the notation for $R_{r_1}(f)$ there, we get

$$\sum_{k=n+1}^{\infty} |c_k| r^k \left(\frac{[k]_q - k}{(q-1)[n]_q} - 1 \right) \leq \sum_{k=n+1}^{\infty} |c_k| r^k \cdot \frac{[k]_q}{(q-1)[n]_q}$$

$$\leq \frac{R_{r_1}(f)}{(q-1)[n]_q} \sum_{k=n+1}^{\infty} \frac{1}{r_1^k} \cdot r^k \cdot q^k$$

$$= \frac{R_{r_1}(f)}{(q-1)[n]_q} \sum_{k=n+1}^{\infty} \left[\left(\frac{r}{r_1} \right)^{1/3} \right]^k \cdot \left[\left(\frac{r}{r_1} \right)^{1/3} \right]^{2k} \cdot q^k$$

$$\leq \frac{R_{r_1}(f)}{(q-1)[n]_q} \cdot \left(\frac{r}{r_1} \right)^{(n+1)/3} \sum_{k=0}^{\infty} \left[\left(\frac{r}{r_1} \right)^{1/3} \right]^k$$

$$= \frac{R_{r_1}(f)}{(q-1)[n]_q} \cdot \left[\left(\frac{r}{r_1} \right)^{1/3} \right]^n \cdot \frac{r^{1/3}}{r_1^{1/3} - r^{1/3}} \leq \frac{R_{r_1}(f)}{(q-1)^3[n]_q^2}$$

$$\leq \frac{1}{(q-1)^3[n]_q^2} \sum_{k=0}^{\infty} |c_k|(k+1)(k+2)^2(r_1 q)^k,$$

where we used the inequalities, $[k]_q \leq k q^k$, $\frac{1}{q^n} \leq \frac{1}{(q-1)[n]_q}$, $\frac{r^{1/3}}{r_1^{1/3} - r^{1/3}} \leq \frac{1}{q-1}$, and where for $\rho = \left(\frac{r}{r_1} \right)^{1/3} \leq \frac{1}{q}$, we used the obvious inequality $\rho^{2k} \cdot q^k \leq 1$.

Collecting now all the estimates and taking into account that $\frac{1}{q-1} + \frac{1}{(q-1)^3} = \frac{q^2 - 2q + 2}{(q-1)^3}$, we arrive at the desired estimate. $\qquad\square$

Next we present a lower approximation estimate.

Theorem 1.8.3 (Gal [58]). *Let $R > q^4 > 1$, $f : \mathbb{D}_R \to \mathbb{C}$ be analytic in $\mathbb{D}_R$, that is, $f(z) = \sum_{k=0}^{\infty} c_k z^k$ for all $z \in \mathbb{D}_R$, and $1 \leq r < \frac{r_1}{q^3} < \frac{R}{q^4}$ be arbitrary*

*fixed. If f is not a polynomial of degree ≤ 1, then for all $n \in \mathbb{N}$ and $|z| \leq r$,
we have*

$$\|L_{n,q}(f) - f\|_r \geq \frac{C_{r,r_1,q}(f)}{[n]_q},$$

where the constant $C_{r,r_1,q}(f)$ depends only on f, r, and r_1. Here $\|f\|_r$ denotes
$\max_{|z| \leq r}\{|f(z)|\}$.

Proof. For $P_q(f)(z)$ defined in the statement of Theorem 1.8.2, all $|z| \leq r$
and $n \in \mathbb{N}$, we have

$$L_{n,q}(f)(z) - f(z) =$$

$$\frac{1}{[n]_q}\left\{ -P_q(f)(z) + \frac{1}{[n]_q}\left[[n]_q^2 \left(L_{n,q}(f)(z) - f(z) + \frac{P_q(f)(z)}{[n]_q} \right) \right] \right\}.$$

In what follows we will apply to this identity the following obvious property:

$$\|F + G\|_r \geq |\, \|F\|_r - \|G\|_r \,| \geq \|F\|_r - \|G\|_r.$$

It follows

$$\|L_{n,q}(f) - f\|_r \geq \frac{1}{[n]_q}\left\{ \|P_q(f)\|_r - \frac{1}{[n]_q}\left[[n]_q^2 \left\| L_{n,q}(f) - f + \frac{P_q(f)}{[n]_q} \right\|_r \right] \right\}.$$

Since by hypothesis f is not a polynomial of degree ≤ 1 in $\mathbb{D}_R$, we get
$\|P_q(f)\|_r > 0$.

Indeed, supposing the contrary it follows that $P_q(f)(z) = 0$ for all $z \in \overline{\mathbb{D}}_r = \{z \in \mathbb{C}; |z| \leq r\}$.

Since simple calculation shows that $P_q(f)(z) = z \cdot \frac{D_q(f)(z) - f'(z)}{q-1}$,
$P_q(f)(z) = 0$ implies $D_q(f)(z) = f'(z)$, for all $z \in \overline{\mathbb{D}}_r \setminus \{0\}$. Taking into
account the representation of f as $f(z) = \sum_{k=0}^{\infty} c_k z^k$, the last equality immediately leads to $c_k = 0$, for all $k \geq 2$, which means that f is linear in $\overline{\mathbb{D}}_r$,
a contradiction with the hypothesis.

Now, by Theorem 1.8.2 we have

$$[n]_q^2 \left\| L_{n,q}(f) - f + \frac{P_q(f)}{[n]_q} \right\|_r \leq Q_{r_1,q}(f),$$

where $Q_{r_1,q}(f)$ is a positive constant depending only on f, r_1, and q.

Since $\frac{1}{[n]_q} \to 0$ as $n \to \infty$, there exists an index n_0 depending only on f,
r, r_1, and q, such that for all $n > n_0$, we have

$$\|P_q(f)\|_r - \frac{1}{[n]_q}\left[[n]_q^2 \left\| L_{n,q}(f) - f + \frac{P_q(f)}{[n]_q} \right\|_r \right] \geq \frac{1}{2}\left\| \frac{P_q(f)}{2} \right\|_r,$$

which immediately implies that

$$\|L_{n,q}(f) - f\|_r \geq \frac{1}{[n]_q} \cdot \frac{1}{2} \|P_q(f)\|_r, \forall n > n_0.$$

For $n \in \{1, \ldots, n_0\}$ we obviously have $\|L_{n,q}(f) - f\|_r \geq \frac{M_{r,r_1,n,q}(f)}{[n]_q}$ with $M_{r,r_1,n,q}(f) = [n]_q \cdot \|L_{n,q}(f) - f\|_r > 0$ (if $\|L_{n,q}(f) - f\|_r$ would be equal to 0, this would imply that f is a linear function, a contradiction).

Therefore, finally we get $\|L_{n,q}(f) - f\|_r \geq \frac{C_{r,r_1,q}(f)}{n}$ for all $n \in \mathbb{N}$, where

$$C_{r,r_1,q}(f) = \min\left\{M_{r,r_1,1,q}(f), \ldots, M_{r,r_1,n_0,q}(f), \frac{1}{2}\|P_q(f)\|_r\right\},$$

which completes the proof. $\qquad\square$

Combining now Theorem 1.8.3 with Theorem 1.8.1, (i), we immediately get the following.

Corollary 1.8.4 (Gal [58]). *Let $R > q^4 > 1$, $f : \mathbb{D}_R \to \mathbb{C}$ be analytic in $\mathbb{D}_R$, that is, $f(z) = \sum_{k=0}^{\infty} c_k z^k$ for all $z \in \mathbb{D}_R$, and $1 \leq r < \frac{r_1}{q^3} < \frac{R}{q^4}$ be arbitrary fixed. If f is not a polynomial of degree ≤ 1, then for all $n \in \mathbb{N}$ we have*

$$\|L_{n,q}(f) - f\|_r \sim \frac{1}{[n]_q},$$

where the constants in the equivalence depend on f, r, r_1, and q but are independent of n.

Concerning the simultaneous approximation, we present the following.

Theorem 1.8.5 (Gal [58]). *Let $R > q^4 > 1$, $f : \mathbb{D}_R \to \mathbb{C}$ be analytic in $\mathbb{D}_R$, that is, $f(z) = \sum_{k=0}^{\infty} c_k z^k$ for all $z \in \mathbb{D}_R$, and $1 \leq r < r^* < \frac{r_1}{q^3} < \frac{R}{q^4}$ be arbitrary fixed. Also, let $p \in \mathbb{N}$. If f is not a polynomial of degree $\leq \max\{1, p-1\}$, then for all $n \in \mathbb{N}$, we have*

$$\|L_{n,q}^{(p)}(f) - f^{(p)}\|_r \sim \frac{1}{[n]_q},$$

where the constants in the equivalence depend on f, r, r^*, r_1, p, and q but are independent of n.

Proof. Since by Theorem 1.8.1, (ii), we have the upper estimate for $\|L_{n,q}^{(p)}(f) - f^{(p)}\|_r$, it remains to prove the lower estimate for $\|L_{n,q}^{(p)}(f) - f^{(p)}\|_r$.

For this purpose, denoting by Γ the circle of radius r^* and center 0, we have the inequality $|v - z| \geq r^* - r$ valid for all $|z| \leq r$ and $v \in \Gamma$. Cauchy's formula is expressed by

$$L_{n,q}^{(p)}(f)(z) - f^{(p)}(z) = \frac{p!}{2\pi i} \int_\Gamma \frac{L_{n,q}(f)(v) - f(v)}{(v - z)^{p+1}} dv.$$

Now, as in the proof of Theorem 1.8.1, (ii), for all $v \in \Gamma$ and $n \in \mathbb{N}$, we have

$$L_{n,q}(f)(v) - f(v) =$$

$$\frac{1}{[n]_q}\left\{-P_q(f)(v) + \frac{1}{[n]_q}\left[[n]_q^2\left(L_{n,q}(f)(v) - f(v) + \frac{P_q(f)(v)}{[n]_q}\right)\right]\right\},$$

which replaced in the above Cauchy's formula implies

$$L_{n,q}^{(p)}(f)(z) - f^{(p)}(z) = \frac{1}{[n]_q}\left\{\frac{p!}{2\pi i}\int_\Gamma -\frac{P_q(f)(v)}{(v-z)^{p+1}}dv\right.$$

$$+\frac{1}{[n]_q}\cdot\frac{p!}{2\pi i}\int_\Gamma \frac{[n]_q^2\left(L_{n,q}(f)(v) - f(v) + \frac{P_q(f)(v)}{[n]_q}\right)}{(v-z)^{p+1}}dv\left.\right\} =$$

$$\frac{1}{[n]_q}\left\{[-P_q(f)(z)]^{(p)}\right.$$

$$+\frac{1}{[n]_q}\cdot\frac{p!}{2\pi i}\int_\Gamma \frac{[n]_q^2\left(L_{n,q}(f)(v) - f(v) + \frac{P_q(f)(v)}{[n]_q}\right)}{(v-z)^{p+1}}dv\left.\right\}.$$

Passing now to $\|\cdot\|_r$, for all $n \in \mathbb{N}$, it follows

$$\|L_{n,q}^{(p)}(f) - f^{(p)}\|_r \geq \frac{1}{[n]_q}\left\{\left\|[-P_q(f)]^{(p)}\right\|_r\right.$$

$$-\frac{1}{[n]_q}\left\|\frac{p!}{2\pi}\int_\Gamma \frac{[n]_q^2\left(L_{n,q}(f)(v) - f(v) + \frac{P_q(f)(v)}{[n]_q}\right)}{(v-z)^{p+1}}dv\right\|_r\left.\right\},$$

where by using Theorem 1.8.2, for all $n \in \mathbb{N}$, we get

$$\left\|\frac{p!}{2\pi}\int_\Gamma \frac{[n]_q^2\left(L_{n,q}(f)(v) - f(v) + \frac{P_q(f)(v)}{[n]_q}\right)}{(v-z)^{p+1}}dv\right\|_r \leq$$

$$\frac{p!}{2\pi}\cdot\frac{2\pi r^*[n]_q^2}{(r^*-r)^{p+1}}\left\|L_{n,q}(f) - f + \frac{P_q(f)}{[n]_q}\right\|_{r^*} \leq Q_{r_1,q}(f)\cdot\frac{p!r^*}{(r^*-r)^{p+1}}.$$

But by hypothesis on f, we have $\left\|-[P_q(f)]^{(p)}\right\|_{r^*} > 0$. Indeed, supposing the contrary, it would follow that $[P_q(f)]^{(p)}(z) = 0$, for all $|z| \leq r^*$, where by the statement of Theorem 1.8.2, we have

$$P_q(f)(z) = \sum_{k=2}^\infty c_k([1]_q + [2]_q + \ldots + [k-1]_q)z^k.$$

Firstly, supposing that $p = 1$, by $P'_q(f)(z) = \sum_{k=2}^{\infty} c_k k([1]_q + [2]_q + \ldots + [k-1]_q)z^{k-1} = 0$, for all $|z| \le r^*$, would follow that $c_k = 0$, for all $k \ge 2$, that is, f would be a polynomial of degree $1 = \max\{1, p-1\}$, a contradiction with the hypothesis.

Taking $p = 2$, we would get $P''_q(z) = \sum_{k=2}^{\infty} c_k k(k-1)([1]_q + [2]_q + \ldots + [k-1]_q)z^{k-2} = 0$, for all $|z| \le r^*$, which immediately would imply that $c_k = 0$, for all $k \ge 2$, that is, f would be a polynomial of degree $1 = \max\{1, p-1\}$, a contradiction with the hypothesis.

Now, taking $p > 2$, for all $|z| \le r^*$, we would get

$$P_q^{(p)}(f)(z) = \sum_{k=p}^{\infty} c_k k(k-1)\ldots(k-p+1)([1]_q + [2]_q + \ldots + [k-1]_q)z^{k-p} = 0,$$

which would imply $c_k = 0$ for all $k \ge p$, that is, f would be a polynomial of degree $p - 1 = \max\{1, p-1\}$, a contradiction with the hypothesis.

In continuation, reasoning exactly as in the proof of Theorem 1.8.3, we immediately get the desired conclusion. $\qquad\square$

Remark. Taking into account that for $q > 1$, we have the inequalities $(q-1) \cdot \frac{1}{q^n} \le \frac{1}{[n]_q} \le q \cdot \frac{1}{q^n}$, for all $n \in \mathbb{N}$, it follows that the exact order of approximation in Corollary 1.8.4 and Theorem 1.8.5 is q^{-n}, which is essentially better than the order of approximation $1/n$, obtained in the case $q = 1$, that is, for $L_{n,1}(f)(z) := L_n(f)(z)$, in [52].

We finish this section by proving some approximation results for the iterates of q-Lorentz polynomials.

For f analytic in $\mathbb{D}_R$ that is of the form $f(z) = \sum_{k=0}^{\infty} c_k z^k$, for all $z \in \mathbb{D}_R$, let us define the iterates of complex Lorentz polynomial $L_{n,q}(f)(z)$, by $L_{n,q}^{(1)}(f)(z) = L_{n,q}(f)(z)$ and $L_{n,q}^{(m)}(f)(z) = L_{n,q}[L_{n,q}^{(m-1)}(f)](z)$, for any $m \in \mathbb{N}$, $m \ge 2$.

Since we have $L_{n,q}(f)(z) = \sum_{k=0}^{\infty} c_k L_{n,q}(e_k)(z)$, by recurrence for all $m \ge 1$, we easily get that $L_{n,q}^{(m)}(f)(z) = \sum_{k=0}^{\infty} c_k L_{n,q}^{(m)}(e_k)(z)$, where $L_{n,q}^{(m)}(e_k)(z) = 1$ if $k = 0$, $L_{n,q}^{(m)}(e_k)(z) = z$ if $k = 1$, $L_{n,q}^{(m)}(e_k)(z) = 0$ if $k \ge n + 1$, and

$$L_{n,q}^{(m)}(e_k)(z) = \left(1 - \frac{[1]_q}{[n]_q}\right)^m \left(1 - \frac{[2]_q}{[n]_q}\right)^m \ldots \left(1 - \frac{[k-1]_q}{[n]_q}\right)^m z^k,$$

for $2 \le k \le n$.

We present the following.

Theorem 1.8.6 (Gal [58]). *Let f be analytic in $\mathbb{D}_R$ with $R > q > 1$, that is, $f(z) = \sum_{k=0}^{\infty} c_k z^k$, for all $z \in \mathbb{D}_R$. Let $1 \le r < \frac{r_1}{q} < \frac{R}{q}$. We have*

$$\|L_{n,q}^{(m)}(f) - f\|_r \le \frac{m}{[n]_q} \cdot \frac{q+1}{(q-1)^2} \sum_{k=0}^{\infty} |c_k|(k+1)r_1^k,$$

and therefore if $\lim_{n \to \infty} \frac{m_n}{[n]_q} = 0$, *then*

$$\lim_{n \to \infty} \|L_{n,q}^{(m_n)}(f) - f\|_r = 0.$$

Proof. For all $|z| \leq r$, we easily obtain

$$|f(z) - L_{n,q}^{(m)}(f)(z)|$$

$$\leq \sum_{k=2}^{n} |c_k| r^k \left[1 - \left(1 - \frac{[1]_q}{[n]_q}\right)^m \left(1 - \frac{[2]}{[n]_q}\right)^m \cdots \left(1 - \frac{[k-1]_q}{[n]_q}\right)^m \right]$$

$$+ \sum_{k=n+1}^{\infty} |c_k| \cdot r^k.$$

Denoting $A_{k,n} = \left(1 - \frac{[1]_q}{[n]_q}\right)\left(1 - \frac{[2]}{[n]_q}\right)\cdots\left(1 - \frac{[k-1]_q}{[n]_q}\right)$, we get $1 - A_{k,n}^m = (1 - A_{k,n})(1 + A_{k,n} + A_{k,n}^2 + \ldots + A_{k,n}^{m-1}) \leq m(1 - A_{k,n})$, and therefore since $1 - A_{k,n} \leq \frac{(k-1)[k-1]_q}{[n]_q}$, for all $|z| \leq r$, we obtain

$$\sum_{k=2}^{n} |c_k| r^k \left[1 - \left(1 - \frac{[1]_q}{[n]_q}\right)^m \left(1 - \frac{[2]_q}{[n]_q}\right)^m \cdots \left(1 - \frac{[k-1]_q}{[n]_q}\right)^m \right]$$

$$\leq m \sum_{k=2}^{\infty} |c_k| r^k [1 - A_{k,n}] \leq \frac{m}{[n]_q} \sum_{k=2}^{\infty} |c_k| (k-1)[k-1]_q r^k$$

$$\leq \frac{m}{[n]_q} \sum_{k=2}^{\infty} |c_k| \cdot \frac{kq^k}{q-1} \cdot r^k \leq \frac{m}{[n]_q} \cdot \frac{1}{q-1} \sum_{k=2}^{\infty} |c_k| (k+1)(rq)^k$$

$$\leq \frac{m}{[n]_q} \cdot \frac{1}{q-1} \sum_{k=2}^{\infty} |c_k| (k+1) r_1^k.$$

On the other hand, following exactly the reasonings in the proof of Theorem 1.8.1, we get the estimate

$$\sum_{k=n+1}^{\infty} |c_k| \cdot r^k \leq \frac{1}{[n]_q} \cdot \frac{2 \sum_{k=0}^{\infty} |c_k| (k+1) r_1^k}{(q-1)^2} \leq \frac{m}{[n]_q} \cdot \frac{2 \sum_{k=0}^{\infty} |c_k| (k+1) r_1^k}{(q-1)^2}.$$

Collecting now all the estimates and taking into account that $\frac{1}{q-1} + \frac{2}{(q-1)^2} = \frac{q+1}{(q-1)^2}$, we arrive at the desired estimate. $\qquad\square$

Remark. Taking into account the equivalence $\frac{1}{[n]_q} \sim \frac{1}{q^n}$, from Theorem 1.8.6, it follows the conclusion that if $\lim_{n \to \infty} \frac{m_n}{q^n} = 0$, then

$$\lim_{n \to \infty} \|L_{n,q}^{(m_n)}(f) - f\|_r = 0.$$

1.9 q-Stancu and q-Stancu–Faber Polynomials, $q \geq 1$

In this section we deal with the approximation properties of the complex q-Stancu polynomials, $q > 1$, attached to analytic functions in compact disks and, as a generalization, with the approximation properties of the complex q-Stancu–Faber polynomials, $q > 1$, attached to analytic functions in compact subsets in $\mathbb{C}$.

Firstly, we present the approximation properties in compact disks for the complex q-Stancu polynomials.

Let $q > 0$. For any $n \in \mathbb{N} \cup \{0\}$, the q-integer $[n]_q$ is defined by

$$[n]_q := 1 + q + \cdots + q^{n-1}, \quad [0]_q := 0;$$

and the q-factorial $[n]_q!$ by

$$[n]_q! := [1]_q [2]_q \cdots [n]_q, \quad [0]_q! := 1.$$

For integers $0 \leq k \leq n$, the q-binomial is defined by

$$\binom{n}{k}_q := \frac{[n]_q!}{[k]_q![n-k]_q!}.$$

The complex q-Stancu polynomials are defined by

$$B_{n,q}^{\beta,\gamma}(f;z) := \sum_{k=0}^{n} f\left(\frac{[k]_q + [\beta]_q}{[n]_q + [\gamma]_q}\right) m_{n,k}(q;z),$$

where $f \in C[0;1], q > 0, 0 \leq \beta \leq \gamma$ and

$$m_{n,k}(q;z) := \binom{n}{k}_q z^k \prod_{s=0}^{n-k-1} (1 - q^s z).$$

We list below the following results due to Mahmudov [104].

Theorem 1.9.1 (Mahmudov [104]). *Let $q \geq 1$ be fixed. If f is analytic in $D_R = \{z \in \mathbb{C}; |z| < R\}$ (i.e., $f(z) = \Sigma_{m=0}^{\infty} a_m z^m$, for all $z \in D_R$) and $R > q^2 \geq q \geq r$, then*

$$|B_{n,q}^{\beta,\gamma}(f;z) - f(z)| \leq \frac{M_{1,q}^{\gamma}(f)}{([n]_q + [\gamma]_q)(q-1)}$$

for $|z| \leq r$ and all $n = 1, 2, \ldots$, where $M_{1,q}^{\gamma}(f) = \Sigma_{m=0}^{\infty} 2m|a_m|(q^{2m} + q^{m+\gamma})$.

Remark. From Theorem 1.9.1, if a function is analytic in a disk of radius $R > q^2$, then for $q > 1$, the rate of approximation by the q-Stancu polynomials is of order q^{-n}. Indeed, this is immediate from the inequalities

$$\frac{q-1}{q^n} \leq \frac{1}{[n]_q} \leq \frac{q}{q^n}.$$

Therefore, in the case when $q > 1$, the approximation of an analytic function with q-Stancu polynomials is essentially better than $1/n$, of the classical complex Stancu polynomials (corresponding to the case $= 1$).

Now, defining the mth iterates by $^mB_{n,q}^{(\beta,\gamma)}$, we have the following result.

Theorem 1.9.2 (Mahmudov [104]). *Let* $q \geq 1$ *and* $D_R = \{z \in \mathbb{C} | z | < R\}$ *be with* $R > 1$. *Suppose that* f *is analytic in* D_R, *i.e.,* $f(z) = \Sigma_{m=0}^{\infty} a_m z^m$, *for all* $z \in D_R$. *Let* $0 \leq \beta \leq \gamma$ *and* $1 \leq r < R$. *Then, for all* $|z| \leq r$, *we have*

$$|^pB_{n,q}^{\beta,\gamma}(f;z) - f(z)| \leq \frac{4p}{[n]_q + [\gamma]_q} \sum_{m=1}^{\infty} |a_m| \left(m[\gamma]_q + \sum_{j=1}^{m} [j-1]_q \right) r^m.$$

Remark. The iterates of $B_{n,q}^{(\beta,\gamma)}$ for $q = 1$ were studied in Sect. 1.7 of the book of Gal [49].

As an immediate consequence we obtain the following.

Corollary 1.9.3 (Mahmudov [104]). *Let* $q \geq 1$. *If* $\frac{m_n}{[n]_q} \to 0$ *for* $n \to \infty$, *then* $^{m_n}B_{n;q}^{(\beta,\gamma)}(f;z) \to f(z)$ *uniformly in* $|z| \leq r$, *for any* $1 \leq r < R$.

Remark. We note that Theorem 1.9.2 and Corollary 1.9.3 are new even for the case of real functions of one real variable.

In what follows we extend Theorem 1.9.1 to the approximation by the more general q-Stancu–Faber polynomials (depending on two parameters $0 \leq \alpha \leq \gamma$) attached to an analytic function on a compact subset G of $\mathbb{C}$. For $q > 1$ and G, a compact disk centered in origin, we recapture Theorem 1.9.1, while for G compact set and $q = 1$, we recapture the result in [49], pp. 19–20, Theorem 1.1.8.

Firstly, let us briefly recall some basic concepts on Faber polynomials and Faber expansions required in the next considerations.

$G \subset \mathbb{C}$ will be considered a compact set such that $\tilde{\mathbb{C}} \setminus G$ is connected. Let $A(G)$ be the Banach space of all functions that are continuous on G and analytic in the interior of G endowed with the uniform norm $\|f\|_G = \sup\{|f(z)|; z \in G\}$. If we denote $\mathbb{D}_r = \{z \in \mathbb{C}; |z| < r\}$, then according to the Riemann mapping theorem, a unique conformal mapping Ψ of $\tilde{\mathbb{C}} \setminus \overline{\mathbb{D}}_1$ onto $\tilde{\mathbb{C}} \setminus G$ exists so that $\Psi(\infty) = \infty$ and $\Psi'(\infty) > 0$. The nth Faber polynomial $F_n(z)$ attached to G can be defined by

$$\frac{\Psi'(w)}{\Psi(w) - z} = \sum_{n=0}^{\infty} \frac{F_n(z)}{w^{n+1}}, \ z \in G, |w| > 1.$$

Then $F_n(z)$ is a polynomial of exact degree n.

If $f \in A(G)$, then

$$a_n(f) = \frac{1}{2\pi i} \int_{|u|=1} \frac{f(\Psi(u))}{u^{n+1}} du = \frac{1}{2\pi} \int_{-\pi}^{\pi} f(\Psi(e^{it})) e^{-int} dt, n \in \mathbb{N} \cup \{0\}$$

(1.9.1)

are called the Faber coefficients of f, and $\sum_{n=0}^{\infty} a_n(f) F_n(z)$ is called the Faber expansion (series) attached to f on G. The Faber series represent a natural generalization of the Taylor series, when the unit disk is replaced by an arbitrary simply connected domain bounded by a "nice" curve.

For further properties of Faber polynomials and Faber expansions see, e.g., Gaier [38] and Suetin [136].

By using the Faber polynomials $F_p(z)$, attached to the compact set G, for $f \in A(G)$, let us introduce the following.

Definition 1.9.4 (Gal [61]). For $0 \leq \alpha \leq \gamma$ and $q \geq 1$, the q-Stancu–Faber polynomials attached to G and f are defined by the formula

$$S_{n,q}^{(\alpha,\gamma)}(f;G)(z) = \sum_{p=0}^{n} \binom{n}{p}_q [\Delta_{1/([n]_q+[\gamma]_q)}^p F(a)]_q \cdot F_p(z), \ z \in G, \ n \in \mathbb{N},$$

where $a = \frac{[\alpha]_q}{[n]_q+[\gamma]_q}$, $F_p(z)$ is the Faber polynomial of degree p attached to G in the Faber's expansion of F,

$$[\Delta_h^p F(a)]_q = \sum_{k=0}^{p} (-1)^k q^{k(k-1)/2} \binom{p}{k}_q F(a + [p-k]_q h),$$

$$F(w) = \frac{1}{2\pi i} \int_{|u|=1} \frac{f(\Psi(u))}{u-w} du = \frac{1}{2\pi} \int_{-\pi}^{\pi} \frac{f(\Psi(e^{it})) e^{it}}{e^{it}-w} dt, \ w \in [0,1], \quad (1.9.2)$$

and Ψ is the conformal mapping of $\tilde{\mathbb{C}} \setminus \overline{\mathbb{D}}_1$ onto $\tilde{\mathbb{C}} \setminus G$.

If $\alpha = \gamma$, then since $F(1)$ is involved in $[\Delta_{1/([n]_q+[\gamma]_q)}^n F(a)]_q$ and therefore in the definition of $S_{n,q}^{(\alpha,\gamma)}(f;G)(z)$ too, in addition we will suppose that F can be extended by continuity on the boundary $\partial \mathbb{D}_1$. Note that if $\alpha < \gamma$, then this additional assumption is not necessary.

Remarks. 1) A sufficient condition for the continuity on $\partial \mathbb{D}_1$ of F in Definition 1.9.4 is, for example, that $\int_0^1 \frac{\omega_p(f \circ \Psi; u)_{\partial \mathbb{D}_1}}{u} du < \infty$ (see, e.g., [38], p. 52, Theorem 6). Here $p \in \mathbb{N}$ is arbitrary fixed.

2) For $G = \overline{\mathbb{D}}_1$ it is easy to see that the above q-Stancu–Faber polynomials reduce to the complex q-Stancu polynomials introduced by Mahmudov [104] and mentioned at the beginning of the section.

Recall that a set G is called a continuum if it is a connected compact subset of $\mathbb{C}$. The function f is called analytic in G, if there exists $R > 1$ such that f is analytic in G_R. Here G_R denotes the interior of the closed level curve Γ_R given by $\Gamma_R = \{\Psi(w); |w| = R\}$, where Ψ is the conformal mapping mentioned above (recall that $G \subset G_R$).

The main result is the following upper estimate.

Theorem 1.9.5 (Gal [61]). *Let $1 < q < R$, $0 \leq \alpha \leq \gamma$ and suppose that f analytic on the continuum G. Also, if $\alpha = \gamma$, then, in addition, suppose that F given in Definition 1.9.4 can be extended by continuity on $\partial \mathbb{D}_1$.*

For any $1 < r < \frac{R}{q}$ the following estimate in approximation by q-Stancu– Faber polynomials

$$|S_{n,q}^{(\alpha,\gamma)}(f;G)(z) - f(z)| \leq \frac{C}{q^n}, \text{ for all } z \in \overline{G}_r, \ n \in \mathbb{N},$$

holds, where $C > 0$ depends on f, r, G_r, α, γ, and q but is independent of n and z.

Proof. First we note that since G is a continuum, then it follows that $\tilde{\mathbb{C}} \setminus G$ is simply connected. By the proof of Theorem 2, p. 52 in [136] (by taking there $K = \overline{G}_r$), for any fixed β satisfying $1 < q < \beta < R$, we have $f(z) = \sum_{k=0}^{\infty} a_k(f)F_k(z)$ uniformly in $\overline{G}_\beta$, where $a_k(f)$ are the Faber coefficients and are given by $a_k(f) = \frac{1}{2\pi i} \int_{|u|=\beta} \frac{f(\Psi(u))}{u^{k+1}} du$. Note here that from the definition of the mapping Ψ and of the set G_β, we immediately have that $G \subset \overline{G}_\beta$.

Firstly, we will prove that

$$S_{n,q}^{(\alpha,\gamma)}(f;G)(z) = \sum_{k=0}^{\infty} a_k(f)S_{n,q}^{(\alpha,\gamma)}(F_k;G)(z), \text{ for all } z \in G.$$

In this sense denote by $f_m(z) = \sum_{k=0}^{m} a_k(f)F_k(z)$, $m \in \mathbb{N}$ the partial sum of order m in the Faber expansion of f.

Since by the linearity of $S_{n,q}^{(\alpha,\gamma)}$, we easily get

$$S_{n,q}^{(\alpha,\gamma)}(f_m;G)(z) = \sum_{k=0}^{m} a_k(f)S_{n,q}^{(\alpha,\gamma)}(F_k;G)(z), \text{ for all } z \in G,$$

it suffices to prove that $\lim_{m\to\infty} S_{n,q}^{(\alpha,\gamma)}(f_m;G)(z) = S_{n,q}^{(\alpha,\gamma)}(f;G)(z)$, for all $z \in G$ and $n \in \mathbb{N}$.

First we have

$$S_{n,q}^{(\alpha,\gamma)}(f_m;G)(z) = \sum_{p=0}^{n} \binom{n}{p}_q [\Delta_{1/([n]_q+[\gamma]_q)}^p G_m(a)]_q F_p(z),$$

where $G_m(w) = \frac{1}{2\pi i} \int_{|u|=1} \frac{f_m(\Psi(u))}{u-w} du$, $F(w) = \frac{1}{2\pi i} \int_{|u|=1} \frac{f(\Psi(u))}{u-w} du$ and a is given by Definition 1.9.4. Note here that since by [38], p. 48, the first relation before (6.17), we have

$$\mathcal{F}_k(w) = \frac{1}{2\pi i} \int_{|u|=1} \frac{F_k(\Psi(u))}{u-w} du = w^k, \text{ for all } |w| < 1,$$

evidently that $\mathcal{F}_k(w)$ can be extended by continuity on $\partial \mathbb{D}_1$. This also immediately implies that $G_m(w) = \frac{1}{2\pi i} \int_{|u|=1} \frac{f_m(\Psi(u))}{u-w} du$ can be extended by continuity on $\partial \mathbb{D}_1$, which means that $\mathcal{S}_{n,q}^{(\alpha,\gamma)}(F_k; G)(z)$ and $\mathcal{S}_{n,q}^{(\alpha,\gamma)}(f_m; G)(z)$ are well defined, for all $0 \leq \alpha \leq \gamma$.

Now, taking into account Cauchy's theorem, we also can write

$$G_m(w) = \frac{1}{2\pi i} \int_{|u|=\beta} \frac{f_m(\Psi(u))}{u-w} du \text{ and } F(w) = \frac{1}{2\pi i} \int_{|u|=\beta} \frac{f(\Psi(u))}{u-w} du.$$

For all $n, m \in \mathbb{N}$ and $z \in G$ it follows

$$|\mathcal{S}_{n,q}^{(\alpha,\gamma)}(f_m; G)(z) - \mathcal{S}_{n,q}^{(\alpha,\gamma)}(f; G)(z)|$$

$$\leq \sum_{p=0}^{n} \binom{n}{p}_q |[\Delta_{1/([n]_q+[\gamma]_q)}^p(G_m - F)(a)]_q| \cdot |F_p(z)|$$

$$\leq \sum_{p=0}^{n} \binom{n}{p}_q \sum_{j=0}^{p} q^{j(j-1)/2} \binom{p}{j}_q |(G_m - F)(a + [p-j]_q/([n]_q + [\gamma]_q))| \cdot |F_p(z)|$$

$$\leq \sum_{p=0}^{n} \binom{n}{p}_q \sum_{j=0}^{p} q^{j(j-1)/2} \binom{p}{j}_q C_{j,p,\beta,\alpha,\gamma} \|f_m - f\|_{\overline{G}_\beta} \cdot |F_p(z)|$$

$$\leq M_{n,p,\beta,G_\beta,q,\alpha,\gamma} \|f_m - f\|_{\overline{G}_\beta},$$

which by $\lim_{m\to\infty} \|f_m - f\|_{\overline{G}_\beta} = 0$ (see, e.g., the proof of Theorem 2, p. 52 in [136]) implies the desired conclusion. Here $\|f_m - f\|_{\overline{G}_\beta}$ denotes the uniform norm of $f_m - f$ on $\overline{G}_\beta$.

Consequently we obtain

$$|\mathcal{S}_{n,q}^{(\alpha,\gamma)}(f; G)(z) - f(z)| \leq \sum_{k=0}^{\infty} |a_k(f)| \cdot |\mathcal{S}_{n,q}^{(\alpha,\gamma)}(F_k; G)(z) - F_k(z)| =$$

$$\sum_{k=0}^{n} |a_k(f)| \cdot |\mathcal{S}_{n,q}^{(\alpha,\gamma)}(F_k; G)(z) - F_k(z)| + \sum_{k=n+1}^{\infty} |a_k(f)| \cdot |\mathcal{S}_{n,q}^{(\alpha,\gamma)}(F_k; G)(z) - F_k(z)|.$$

Therefore it remains to estimate $|a_k(f)| \cdot |\mathcal{S}_{n,q}^{(\alpha,\gamma)}(F_k;G)(z) - F_k(z)|$, firstly for all $0 \leq k \leq n$ and secondly for $k \geq n+1$, where

$$\mathcal{S}_{n,q}^{(\alpha,\gamma)}(F_k;G)(z) = \sum_{p=0}^{n} \binom{n}{p}_q [\Delta_{1/([n]_q+[\gamma]_q)}^p F_k(a)]_q \cdot F_p(z).$$

It is useful to observe that by [38], p. 48, the first relation before (6.17), combined with Cauchy's theorem, for any fixed $q < \beta < R$, we have

$$\mathcal{F}_k(w) := \frac{1}{2\pi i} \int_{|u|=\beta} \frac{F_k(\Psi(u))}{u-w} du = w^k = e_k(w), \text{ for all } |w| < \beta.$$

Denote

$$D_{n,p,k}^{(q)}(\alpha,\gamma) = \binom{n}{p}_q [\Delta_{1/([n]_q+[\gamma]_q)}^p e_k(a)]_q.$$

By the proof of Lemma 1, p. 3765 in Mahmudov [104], we can write

$$D_{n,p,k}^{(q)}(\alpha,\gamma) = \left(1 - \frac{[\gamma]_q}{[n]_q+[\gamma]_q}\right) \cdots \left(1 - \frac{[p-1]_q+[\gamma]_q}{[n]_q+[\gamma]_q}\right)$$

$$\cdot [[\alpha]_q/([n]_q+[\gamma]_q), \ldots, ([p]_q+[\alpha]_q)/([n]_q+[\gamma]_q); e_k],$$

where $[y_0, y_1, \ldots, y_p; e_k]$ denotes the divided difference of $e_k(z) = z^k$ on the points $y_0, \ldots, y_p$.

It follows

$$\mathcal{S}_{n,q}^{(\alpha,\gamma)}(F_k;G)(z) = \sum_{p=0}^{n} D_{n,p,k}^{(q)}(\alpha,\gamma) \cdot F_p(z).$$

Since each e_k is convex of any order, by Lemma 2, p. 3765 in Mahmudov [104], it follows that all $D_{n,p,k}^{(q)}(\alpha,\gamma) \geq 0$ and $\sum_{p=0}^{n} D_{n,p,k}^{(q)}(\alpha,\gamma) = \left(\frac{[n]_q+[\alpha]_q}{[n]_q+[\gamma]_q}\right)^k \leq 1$, for all k and n.

Also, note that for all $k \geq 1$,

$$D_{n,k,k}^{(q)}(\alpha,\gamma) = \left(1 - \frac{[\gamma]_q}{[n]_q}\right) \cdots \left(1 - \frac{[k-1]_q+[\gamma]_q}{[n]_q}\right)$$

and that $D_{n,0,0}^{(q)}(\alpha,\gamma) = 1$.

In the estimation of $|a_k(f)| \cdot |\mathcal{S}_{n,q}^{(\alpha,\gamma)}(F_k;G)(z) - F_k(z)|$, we distinguish two cases: 1) $0 \leq k \leq n$; 2) $k > n$.

Case 1. Since $D_{n,0,0}^{(q)}(\alpha,\gamma) = 1$, we may suppose that $1 \leq k \leq n$. We have

$$|\mathcal{S}_{n,q}^{(\alpha,\gamma)}(F_k;G)(z) - F_k(z)| \leq |F_k(z)| \cdot |1 - D_{n,k,k}^{(q)}| + \sum_{p=0}^{k-1} D_{n,p,k}^{(q)}(\alpha,\gamma) \cdot |F_p(z)|.$$

Fix now $1 < r < \frac{\beta}{q}$. By the inequality (8), p. 43 in [136] we have

$$|F_p(z)| \leq C(r)r^p, \text{ for all } z \in \overline{G}_r, p \geq 0.$$

Indeed, by relationship (8) above mentioned (with r instead of R there), we have $|F_p(z)| \leq C(r)r^p$, for all $z \in \Gamma_r$, which, by the Maximum Modulus Theorem for analytic functions, implies $|F_p(z)| \leq c(r)r^p$ for all $z \in \overline{G}_r$ (for these estimates, see also Curtiss [29], page 583, relationship (4.1) and the next two lines).

It is also worth noting that similar estimates hold from page 42, relationships (1), (3), and (4) in Suetin[136], by taking there $r = 1 + \varepsilon$ and $K = \overline{G}_{r'}$, with $1 < r' < r$ arbitrary close to r (in this case we get $|F_p(z)| \leq C(r)r^p$, for all $z \in \overline{G}_{r'}, p \geq 0$, but which still is good enough for the proof, because r in $1 < r < R$ is arbitrary and $r' < r$ is arbitrary close to r).

Now, by the proof of Theorem 3, p. 3766 in Mahmudov [104], we immediately get

$$|\mathcal{S}_{n,q}^{(\alpha,\gamma)}(F_k;G)(z) - F_k(z)| \leq 2C(r)[1 - D_{n,k,k}^{(q)}]r^k \leq c(r)\frac{k[k-1]_q + k[\gamma]_q}{[n]_q + [\gamma]_q}r^k$$

$$\leq c(r)\frac{k(q^k + q^\gamma)}{(q-1)([n]_q + [\gamma]_q)}r^k = c(r)\frac{k(qr)^k}{(q-1)([n]_q + [\gamma]_q)}$$

$$+c(r)q^\gamma\frac{kr^k}{([n]_q + [\gamma]_q)(q-1)} \leq c(r,q,\gamma)\frac{k(qr)^k}{[n]_q},$$

for all $z \in \overline{G}_r$.

Also by the above formula for $a_k(f)$, we easily obtain $|a_k(f)| \leq \frac{C(\beta,f)}{\beta^k}$, for all $k \geq 0$. Note that $C(r), c(r), c(r,q,\gamma), C(\beta,f) > 0$ are positive constants independent of k.

For all $z \in \overline{G}_r$ and $k = 0, 1, 2, \ldots n$, it follows

$$|a_k(f)| \cdot |\mathcal{S}_{n,q}^{(\alpha,\gamma)}(F_k;G)(z) - F_k(z)| \leq \frac{C(r,\beta,f,q,\gamma)}{[n]_q}k\left[\frac{qr}{\beta}\right]^k,$$

that is,

$$\sum_{k=0}^{n}|a_k(f)| \cdot |\mathcal{S}_{n,q}^{(\alpha,\gamma)}(F_k;G)(z) - F_k(z)| \leq \frac{C(r,\beta,f,q,\gamma)}{[n]_q}\sum_{k=1}^{n}kd_1^k, \text{ for all } z \in \overline{G}_r,$$

where $d_1 = \frac{rq}{\beta} < 1$.

Also, clearly we have $\sum_{k=1}^{n} kd_1^k \leq \sum_{k=1}^{\infty} kd_1^k < \infty$ which finally implies that

$$\sum_{k=0}^{n} |a_k(f)| \cdot |\mathcal{S}_{n,q}^{(\alpha,\gamma)}(F_k;G)(z) - F_k(z)| \leq \frac{C^*(r,\beta,f,q,\gamma)}{[n]_q} \leq \frac{qC^*(r,\beta,f,q,\gamma)}{q^n}.$$

Here we used the inequality $\frac{1}{[n]_q} \leq \frac{q}{q^n}$.

Case 2. We have

$$\sum_{k=n+1}^{\infty} |a_k(f)| \cdot |\mathcal{S}_{n,q}^{(\alpha,\gamma)}(F_k;G)(z) - F_k(z)| \leq \sum_{k=n+1}^{\infty} |a_k(f)| \cdot |\mathcal{S}_{n,q}^{(\alpha,\gamma)}(F_k;G)(z)| +$$

$$\sum_{k=n+1}^{\infty} |a_k(f)| \cdot |F_k(z)|.$$

By the estimates mentioned in Case 1, we immediately get

$$\sum_{k=n+1}^{\infty} |a_k(f)| \cdot |F_k(z)| \leq C(r,\beta,f) \sum_{k=n+1}^{\infty} d^k, \text{ for all } z \in \overline{G}_r,$$

with $d = r/\beta$.

Also,

$$\sum_{k=n+1}^{\infty} |a_k(f)| \cdot |\mathcal{S}_{n,q}^{(\alpha,\gamma)}(F_k;G)(z)| = \sum_{k=n+1}^{\infty} |a_k(f)| \cdot \left| \sum_{p=0}^{n} D_{n,p,k}^{(q)}(\alpha,\gamma) \cdot F_p(z) \right|$$

$$\leq \sum_{k=n+1}^{\infty} |a_k(f)| \cdot \sum_{p=0}^{n} D_{n,p,k}^{(q)}(\alpha,\gamma) \cdot |F_p(z)|.$$

But for $p \leq n < k$ and taking into account the estimates obtained in Case 1, we get

$$|a_k(f)| \cdot |F_p(z)| \leq C(r,\beta,f)\frac{r^p}{\beta^k} \leq C(r,\beta,f)\frac{r^k}{\beta^k}, \text{ for all } z \in \overline{G}_r,$$

which therefore implies

$$\sum_{k=n+1}^{\infty} |a_k(f)| \cdot |\mathcal{S}_{n,q}^{(\alpha,\gamma)}(F_k;G)(z) - F_k(z)|$$

$$\leq C(r,\beta,f) \sum_{k=n+1}^{\infty} \sum_{p=0}^{n} D_{n,p,k}^{(q)}(\alpha,\gamma) \left[\frac{r}{\beta}\right]^k + C(r,\beta,f) \sum_{k=n+1}^{\infty} \left[\frac{r}{\beta}\right]^k$$

$$\leq C(r,\beta,f,\gamma) \sum_{k=n+1}^{\infty} \left[\frac{r}{\beta}\right]^k = C(r,\beta,f,\gamma)\frac{d^{n+1}}{1-d}$$

$$= \frac{rC(r,\beta,f,\gamma)}{\beta-r} \cdot d^n \leq \frac{rC(r,\beta,f,\gamma)}{\beta-r} \cdot \frac{1}{q^n},$$

with $d = \frac{r}{\beta} < \frac{1}{q} < 1$.

In conclusion, collecting the estimates in Cases 1 and 2, we obtain

$$|S_{n,q}^{(\alpha,\gamma)}(f;G)(z) - f(z)| \leq \frac{c_1}{q^n} + \frac{c_2}{q^n} \leq \frac{C}{q^n}, \ z \in \overline{G}_r, \ n \in \mathbb{N},$$

with the constants $c_1, c_2, C > 0$ depending on r, β, f, q, γ but independent of n and z. This proves the theorem. □

Remarks. 1) For $\alpha < \gamma$, the q-Stancu–Faber polynomials present the advantage that they do not require the additional condition concerning the extension by continuity of F on the boundary $\partial \mathbb{D}_1$.
2) Since for each $r > 1$, we clearly have $G \subset \overline{G}_r$, obviously the upper estimate in Theorem 1.9.5 holds on the compact set G too.

At the end of this section we just summarize a list of potential applications of the Theorem 1.9.5 to several particular cases of G, when the conformal mapping Ψ and the associated Faber polynomials $F_n(z)$ can explicitly be obtained and so the construction of q-Stancu–Faber polynomials is possible.

Application 1.9.6 (Gal [61]). For $G = \overline{\mathbb{D}}_r$ with $r > 1$, it is well known that $\Psi(z) = z$, $F(z) = f(z)$, $F_p(z) = z^p$ (see, e.g., Stepanets [135], p. 36) and therefore $S_{n,q}^{(\alpha,\gamma)}(f;G)(z)$ become the classical q-Stancu polynomials, while Theorem 1.9.5 becomes the upper approximation estimate in Theorem 1.9.1.

Application 1.9.7 (Gal [61]). Suppose now $G = [-1,1]$. It is well known that $\Psi(w) = \frac{1}{2}\left(w + \frac{1}{w}\right)$ (see, e.g., [97], p. 42), $F_p(z) = 2\cos[p \cdot \arccos z]$ (see, e.g., [77], p. 307). In this case, by the formulas $\cos(t) = \frac{e^{it} + e^{-it}}{2}$, $\cot(t) = \frac{\cos t}{\sin t}$, we get

$$F(w) = \frac{1}{2\pi}\int_{-\pi}^{\pi} f(\cos t) \cdot \frac{e^{it}}{e^{it} - w}dt = \frac{1}{2\pi}\int_{-\pi}^{\pi} f(\cos t) \cdot \frac{1 - w\cos t}{1 - 2w\cos t + w^2}dt$$

$$- \frac{wi}{2\pi}\int_{-\pi}^{\pi} f(\cos t) \cdot \frac{\sin t}{1 - 2w\cos t + w^2}dt, \ w \in [0,1),$$

$$F(1) = \frac{1}{4\pi}\int_{-\pi}^{\pi} f(\cos t)dt - \frac{i}{4\pi}\int_{-\pi}^{\pi} f(\cos t)\cot(t/2)dt,$$

and $S_{n,q}^{(\alpha,\gamma)}(f;[-1,1])(z)$ given by the Definition 1.9.4 could be called q-Stancu–Chebyshev polynomials.

Note that above we have $-\frac{i}{4\pi}\int_{-\pi}^{\pi} f(\cos t)\cot(t/2)dt = \frac{i}{2}\tilde{g}(0)$ with $g(t) = f(\cos t)$, where $\tilde{g}$ is the so-called conjugate of the 2π-periodic function g, defined by (see, e.g., [135], p. 21)

$$\tilde{g}(x) = -\frac{1}{2\pi}\int_{-\pi}^{\pi} g(x+t)\cot(t/2)dt,$$

with the singular integral (at $t = 0$) considered in the sense of principal value. Therefore we can write

$$F(1) = \frac{1}{4\pi}\int_{-\pi}^{\pi} f(\cos t)dt + \frac{i}{2}\widetilde{f(\cos)}(0).$$

And if f is analytic in an open set containing $[-1,1]$ (e.g., in an ellipse with foci at -1 and 1), then by, e.g., Theorem 5.2 in [135], p. 21, $\widetilde{f(\cos)}(x)$ is well defined (finite) at any $x \in \mathbb{R}$, which implies that $F(1)$ is well defined.

Note that in order for $F(1)$ to exist , the analyticity of f can be replaced with the condition that f is continuous differentiable on $\mathbb{R}$ and that $f(1) = 0$. Indeed, by L'Hospital's rule in this case, it is easy to see that the function $f(\cos t)\cot(t/2)$ can be extended by continuity at the singular point $t = 0$. In conclusion, from Theorem 1.9.5 it follows that the approximation order by the q-Stancu–Chebyshev polynomials $S_{n,q}^{(\alpha,\gamma)}(f;G)(z)$, in any disk G containing $[-1,1]$, is q^{-n}.

Application 1.9.8 (Gal [61]). Let G be bounded by the m-cusped hypocycloid H_m, $m = 2, 3, \dots$, given by the parametric equation

$$z = e^{i\theta} + \frac{1}{m-1}e^{-(m-1)i\theta}, \ \theta \in [0, 2\pi).$$

It is known that the conformal mapping is given by $\Psi(w) = w + \frac{1}{(m-1)w^{m-1}}$ (see, e.g., [82], Proposition 2.1) and that the associated Faber polynomials can be explicitly calculated (see, e.g., [82], Proposition 2.3, and [78]). Without to enter into details, in this case, in the construction of the q-Stancu–Faber polynomials, one relies on the calculation of the integrals in (1.9.2).

Application 1.9.9 (Gal [61]). Other concrete cases of sets G when the Faber polynomials can be explicitly calculated and we could construct the q-Stancu–Faber polynomials are the following:

a) G is the regular m-star ($m = 2, 3, \dots,$) given by

$$S_m = \{x\omega^k; 0 \le x \le 4^{1/m}, k = 0, 1, \dots, m-1, \omega^m = 1\},$$

$\Psi(w) = w\left(1 + \frac{1}{w^m}\right)^{2/m}$ (see, e.g., [81], p. 395) and the Faber polynomials can explicitly be calculated as in, e.g., [12], p. 279.

b) G is the m-leafed symmetric lemniscate, $m = 2, 3, \dots$, with its boundary given by

$$L_m = \{z \in \mathbb{C}; |z^m - 1| = 1\},$$

$\Psi(w) = w \left(1 + \frac{1}{w^m}\right)^{1/m}$ and the corresponding Faber polynomials can explicitly be calculated as in, e.g., [79].

c) G is the semidisk

$$SD = \{z \in \mathbb{C}; |z| \leq 1 \text{ and } |Arg(z)| \leq \pi/2\},$$

$\Psi(w) = \frac{2(w^3-1)+3(w^2-w)+2(w^2+w+1)^{3/2}}{w(w+1)\sqrt{3}}$ (see formula (10), p. 235 in [27])

and the attached Faber polynomials can be calculated as in [27].

d) G is a circular lune (see, e.g., [80]) or G is an annulus sector (see, e.g., [28]), cases when the conformal mapping Ψ and the Faber polynomials can explicitly be calculated.

Remark. With the above notations, denoting the partial sum of order n of the Faber expansion by $P_n(f; G)(z) = \sum_{k=0}^{n} a_k(f)F_k(z)$, one sees that the coefficients $a_k(f)$ are calculated by the integrals in (1.9.1), while the construction of the q-Stancu–Faber polynomials, $S_{n,q}^{(\alpha,\gamma)}(f; G)(z)$, is mainly based on the calculation of the integrals in (1.9.2). Therefore, based on the above concrete examples too, we can conclude that for $q > 1$, the approximation by the q-Stancu–Faber polynomials attached to a function defined on a compact subset of the complex plane and given by Definition 1.9.4 can represent a good alternative to the approximation by the partial sums of the Faber series attached to the same function and subset, both giving a geometric progression order of approximation.

1.10 q-Favard–Szász–Mirakjan Operators, $q > 1$

In this section we list without proofs the approximation results for the complex q-Favard–Szász–Mirakjan operators, $q > 1$, due to Mahmudov [105].

Let us keep the notations for $[n]_q$, $[n]_q!$, $\binom{n}{k}_q$, $D_q(f)(z)$ and $D_q^k(f)(z)$ in the beginning of Sect. 1.9, and in addition, for $q > 1$, let us define the q-exponential by

$$exp_q(z) = \sum_{k=0}^{\infty} \frac{z^k}{[n]_q!} = \Pi_{j=0}^{\infty}\left(1 + (q-1)\frac{z}{q^{j+1}}\right), z \in \mathbb{C}$$

and the complex q-Favard–Szász–Mirakjan operator attached to $f : [0, \infty) \to \mathbb{C}$ given by (see Mahmudov [105])

$$F_{n,q}(f)(z) = \sum_{k=0}^{\infty} f\left(\frac{[k]_q}{[n]_q}\right) \cdot \frac{1}{q^{k(k-1)/2}} \cdot \frac{[n]_q^k z^k}{[k]_q!} \cdot exp_q(-[n]_q q^{-k} z).$$

If f is bounded on $[0, \infty)$, then it is clear that $F_{n,q}(f)(z)$ is well defined for all $z \in \mathbb{C}$. Also, note that in the case when $q \searrow 1$, these operators coincide with the classical complex Favard–Szász–Mirakjan operators, already studied in the book of Gal [49], Section 1.8, pp. 114–119, Theorems 1.8.4–1.8.8 (see also the paper of Gal [59]).

The first main result refers to an upper approximation estimate.

Theorem 1.10.1 (Mahmudov [105]). *For* $1 < q < \frac{R}{2}$, *let* $f : \overline{\mathbb{D}_R} \bigcup [R, \infty) \to \mathbb{C}$ *be analytic in* $\mathbb{D}_R$, *that is,* $f(z) = \sum_{k=0}^{\infty} c_k z^k$, *for all* $|z| < R$, *and continuous and bounded in* $\overline{\mathbb{D}_R} \bigcup [R, \infty)$. *If* $1 \le r < \frac{R}{2q}$ *is fixed, then for all* $|z| \le r$ *and* $n \in \mathbb{N}$, *we have*

$$|F_{n,q}(f)(z) - f(z)| \le \frac{C_{r,q}}{[n]_q},$$

where $C_{r,q} = 2\sum_{k=2}^{\infty} |c_k|(k-1)(2qr)^{k-1} < \infty$.

The following is a Voronovskaja-type result with a quantitative estimate.

Theorem 1.10.2. *[Mahmudov [105]] For* $1 < q < \frac{R}{2}$, *let* $f : \overline{\mathbb{D}_R} \bigcup [R, \infty) \to \mathbb{C}$ *be analytic in* $\mathbb{D}_R$, *that is,* $f(z) = \sum_{k=0}^{\infty} c_k z^k$, *for all* $|z| < R$, *and continuous and bounded in* $\overline{\mathbb{D}_R} \bigcup [R, \infty)$. *If* $1 \le r < \frac{R}{2q}$ *is fixed, then for all* $|z| \le r$ *and* $n \in \mathbb{N}$, *we have*

$$\left| F_{n,q}(f)(z) - f(z) - \frac{L_q(f)(z)}{[n]_q} \right| \le \frac{A_{r,q}|z|}{[n]_q^2},$$

where

$$L_q(f)(z) = \sum_{m=2}^{\infty} c_m \cdot \frac{[m]_q - m}{q - 1} \cdot z^{m-1} = \sum_{m=2}^{\infty} c_m([1]_q + \ldots + [m-1]_q)z^{m-1}$$

and $A_{r,q} = 4\sum_{m=3}^{\infty} |c_m|(m-1)(m-2)(2qr)^{m-3} < \infty$.

The following lower estimate holds.

Theorem 1.10.3 (Mahmudov [105]). *For* $1 < q < \frac{R}{2}$ *and* $1 \le r < \frac{R}{2q}$, *let* $f : \overline{\mathbb{D}_R} \bigcup [R, \infty) \to \mathbb{C}$ *be analytic in* $\mathbb{D}_R$, *that is,* $f(z) = \sum_{k=0}^{\infty} c_k z^k$, *for all* $|z| < R$, *and continuous and bounded in* $\overline{\mathbb{D}_R} \bigcup [R, \infty)$. *If* f *is not a polynomial of degree* ≤ 1, *then for all* $n \in \mathbb{N}$, *we have*

$$\|F_{n,q}(f) - f\|_r \ge \frac{C_{r,q}(f)}{[n]_q},$$

where the constant $C_{r,q}(f)$ *is independent of* n *but depends on* f, r, *and* q. *Here recall that* $\|f\|_r = \max\{|f(z)|; |z| \le r\}$.

Combining Theorem 1.10.1 with Theorem 1.10.3, we immediately get the following.

Theorem 1.10.4 (Mahmudov [105]). *For $1 < q < \frac{R}{2}$ and $1 \leq r < \frac{R}{2q}$, let*
$f : \overline{\mathbb{D}_R} \bigcup [R, \infty) \to \mathbb{C}$ *be analytic in $\mathbb{D}_R$, that is, $f(z) = \sum_{k=0}^{\infty} c_k z^k$, for all*
$|z| < R$, *and continuous and bounded in $\overline{\mathbb{D}_R} \bigcup [R, \infty)$. If f is not a polynomial*
of degree ≤ 1 then for all $n \in \mathbb{N}$, we have

$$\|F_{n,q}(f) - f\|_r \sim \frac{1}{[n]_q},$$

where the constants in the equivalence depend on f, r, and q but are inde-
pendent of n.

Remark. The ideas of the proofs in the approximation results by complex
q-Favard–Szász–Mirakjan operators, without exponential growth imposed to
the approximated function f, appear for the first time in the case $q = 1$ in
the book of Gal [49], pp. 114–119, Theorems 1.8.4–1.8.8. Consequently, the
above results in Mahmudov [105] are generalizations of the results in Gal [49],
pp. 114–119, Theorems 1.8.4–1.8.8.

1.11 q-Bernstein–Faber-Type Polynomials, $q \geq 1$

In my recent book of Gal [49], Section 1.1.2, pp. 19–25, I introduced the
so-called Bernstein–Faber polynomials attached to a compact subset in $\mathbb{C}$
and studied their approximation properties.

In this section, attaching to a compact set $G \subset \mathbb{C}$ and to an analytic function on G the so-called q-Bernstein–Faber polynomials, $q > 1$, for the approximation by these polynomials, quantitative (upper and exact) estimates
of order q^{-n} in G are obtained.

These estimates essentially improve the approximation order $\frac{1}{n}$ from the
approximation by Bernstein–Faber polynomials (corresponding in fact to the
case $q = 1$) in compact subset of $\mathbb{C}$ obtained in the above-mentioned book.

Also, Voronovskaja-type results with quantitative estimates are obtained
for $q \geq 1$.

In the case when $G = \{z \in \mathbb{C}; |z| \leq r\}$, $r \geq 1$, one recaptures the results
in Ostrovska [117] and [145]. Also, for other particular choices of G, several
potential applications will be indicated.

Firstly let us briefly recall some classical concepts and results about Faber
polynomials and Faber expansions.

If $G \subset \mathbb{C}$ is a compact set such that $\tilde{\mathbb{C}} \backslash G$ is connected, denote by $A(G)$ the
Banach space of all functions that are continuous on G and analytic in the
interior of G, endowed with the uniform norm $\|f\|_G = \sup\{|f(z)|; z \in G\}$. If
we denote $\mathbb{D}_r = \{z \in \mathbb{C}; |z| < r\}$, then according to the Riemann mapping
theorem, a unique conformal mapping Ψ of $\tilde{\mathbb{C}} \backslash \overline{\mathbb{D}}_1$ onto $\tilde{\mathbb{C}} \backslash G$ exists so that
$\Psi(\infty) = \infty$ and $\Psi'(\infty) > 0$. The nth *Faber polynomial* $F_n(z)$ attached to G
may be defined by

$$\frac{\Psi'(w)}{\Psi(w) - z} = \sum_{n=0}^{\infty} \frac{F_n(z)}{w^{n+1}}, \ z \in G, |w| > 1.$$

Then $F_n(z)$ is a polynomial of exact degree n.

If $f \in A(\overline{G})$, then

$$a_n(f) = \frac{1}{2\pi i} \int_{|u|=1} \frac{f[\Psi(u)]}{u^{n+1}} du = \frac{1}{2\pi} \int_{-\pi}^{\pi} f[\Psi(e^{it})] e^{-int} dt, n \in \mathbb{N} \cup \{0\}$$

(1.11.1)

are called the Faber coefficients of f, and $\sum_{n=0}^{\infty} a_n(f) F_n(z)$ is called the Faber expansion (series) attached to f on G. (Here $i^2 = -1$.) The Faber series represent a natural generalization of Taylor series when the unit disk is replaced by an arbitrary simply connected domain bounded by a "nice" curve.

For further properties of Faber polynomials and Faber expansions see, e.g., Gaier [38] and Suetin [136].

Firstly, by using the Faber polynomials $F_p(z)$, attached to the compact set G, for $f \in A(G)$, let us

Taking $\alpha = \gamma = 0$ in our Definition 1.9.4 in Sect. 1.9, we get the following.

Definition 1.11.1. For $q \geq 1$, the q-Bernstein–Faber polynomials attached to G and f are defined by the formula

$$\mathcal{B}_{n,q}(f; G)(z) = \sum_{p=0}^{n} \binom{n}{p}_q [\Delta_{1/[n]_q}^p F(0)]_q \cdot F_p(z), \ z \in G, \ n \in \mathbb{N},$$

where $F_p(z)$ is the Faber polynomial of degree p,

$$[\Delta_h^p F(0)]_q = \sum_{k=0}^{p} (-1)^k q^{k(k-1)/2} \binom{p}{k}_q F([p-k]_q h),$$

$$F(w) = \frac{1}{2\pi i} \int_{|u|=1} \frac{f(\Psi(u))}{u - w} du = \frac{1}{2\pi} \int_{-\pi}^{\pi} \frac{f(\Psi(e^{it})) e^{it}}{e^{it} - w} dt, \ w \in [0, 1],$$

and Ψ is the conformal mapping of $\tilde{\mathbb{C}} \setminus \overline{\mathbb{D}}_1$ onto $\tilde{\mathbb{C}} \setminus G$.

Here, since $F(1)$ is involved in $[\Delta_{1/[n]_q}^n F(0)]_q$ and therefore in the definition of $\mathcal{B}_{n,q}(f; G)(z)$ too, in addition we will suppose that F can be extended by continuity on the boundary $\partial \mathbb{D}_1$.

Remarks. 1) Recall that in Definition 1.11.1, we have $[n]_q = \frac{q^n - 1}{q - 1}$ for $q \neq 1$, $[n]_1 = n$,

$$[n]_q! = [1]_q [2]_q \cdot \ldots \cdot [n]_q \text{ and } \binom{n}{k}_q = \frac{[n]_q!}{[k]_q! [n-k]_q!}.$$

2) A sufficient condition for the continuity on $\partial \mathbb{D}_1$ of F in Definition 1.11.1 is, for example, that $\int_0^1 \frac{\omega_p(f \circ \Psi; u)_{\partial \mathbb{D}_1}}{u} du < \infty$ (see, e.g., Gaier [38], p. 52, Theorem 6). Here $p \in \mathbb{N}$ is arbitrary fixed.

3) For $G = \overline{\mathbb{D}}_1$, it is easy to see that in the above q-Bernstein–Faber polynomials, one reduces to the complex q-Bernstein polynomials introduced in the case of real variable in Phillips [123] and given by

$$B_{n,q}(f)(z) = \sum_{p=0}^{n} \binom{n}{p}_q [\Delta^p_{1/[n]_q} f(0)]_q z^p.$$

4) For $q = 1$ in Definition 1.11.1 we recapture the Bernstein–Faber polynomials introduced and studied in Gal [49], pp. 19–25.

In the first main results one refers to the upper estimate in approximation by the q-Bernstein–Faber polynomials, $q \geq 1$, introduced by Definition 1.11.1, on compact sets without any restriction on their boundaries and can be stated as follows. Thus, taking $\alpha = \gamma = 0$ in the statement of our Theorem 1.9.5 in Sect. 1.9, we get the following upper estimate.

Theorem 1.11.2. *Let $q > 1$ and G be a continuum (i.e., a connected compact subset of $\mathbb{C}$). Suppose that f is analytic in G, namely, that there exists $R > 1$ such that f is analytic in G_R. Here recall that G_R denotes the interior of the closed level curve Γ_R given by $\Gamma_R = \{\Psi(w); |w| = R\}$. Also, we suppose that F given in Definition 1.11.1 can be extended by continuity on $\partial \mathbb{D}_1$.*

Let $1 < q < R$. For any $1 < r < \frac{R}{q}$, the following estimate in approximation by q-Bernstein–Faber polynomials

$$|\mathcal{B}_{n,q}(f; G)(z) - f(z)| \leq \frac{C}{q^n}, \text{ for all } z \in \overline{G}_r, \ n \in \mathbb{N},$$

holds, where $C > 0$ depends on f, r, G_r, and q but is independent of n.

Also, recall the following result in the case $q = 1$.

Theorem 1.11.3 (Gal [49], pp. 19–20, Theorem 1.1.8). *Let G be a continuum (i.e., a connected compact subset of $\mathbb{C}$) and suppose that f is analytic in G, that is, there exists $R > 1$ such that f is analytic in G_R. Here recall that G_R denotes the interior of the closed level curve Γ_R given by $\Gamma_R = \{\Psi(w); |w| = R\}$ (and that $G \subset \overline{G}_r$ for all $1 < r < R$). Also, we suppose that F given in the definition of Bernstein–Faber polynomials can be extended by continuity on $\partial \mathbb{D}_1$.*

For any $1 < r < R$ the following upper estimate

$$|\mathcal{B}_{n,1}(f; G)(z) - f(z)| \leq \frac{C}{n}, \ z \in \overline{G}_r, \ n \in \mathbb{N},$$

holds, where $C > 0$ depends on f, r, and G_r but it is independent of n.

Remark. It is clear that the supposition in the Theorems 1.11.2 and 1.11.3 that F given in Definition 1.11.1 can be extended by continuity on $\partial\mathbb{D}_1$ could be weakened to the fact that F can be extended by continuity at the point 1 only.

Now, in order to obtain exact orders of approximation by q-Bernstein–Faber polynomials, $q \geq 1$, we would need to prove only the corresponding lower estimates.

For that purpose, we will use a Voronovskaja-type result with quantitative upper estimates, valid for all $q \geq 1$ and given by the following.

Theorem 1.11.4. *Let $G \subset \mathbb{C}$ be a continuum and suppose that f is analytic in G, that is, there exists $R > 1$ such that f is analytic in G_R, with $f(z) = \sum_{k=0}^{\infty} a_k(f)F_k(z)$ for all $z \in G_R$. In addition, suppose that F given in the Definition 1.11.1 of q-Bernstein–Faber polynomials can be extended by continuity on $\partial\mathbb{D}_1$.*

Assume that $1 \leq q < R$ and denote $S_k^{(q)} = [1]_q + \ldots + [k-1]_q$, $k \geq 2$:

(i) If $q = 1$, then for any $1 < r < R$, $z \in \overline{G_r}$ and $n \in \mathbb{N}$, the following upper estimate

$$\left| \mathcal{B}_{n,1}(f;G)(z) - f(z) - \sum_{k=2}^{\infty} a_k(f) \cdot \frac{S_k^{(1)}}{n}[F_{k-1}(z) - F_k(z)] \right| \leq \frac{C}{n^2}$$

holds, where $C > 0$ depends on f, r, G_r but it is independent of n.

(ii) If $q > 1$, $1 < r < \frac{R}{q^2}$, $z \in \overline{G_r}$ and $n \in \mathbb{N}$, then the following upper estimate

$$\left| \mathcal{B}_{n,q}(f;G)(z) - f(z) - \sum_{k=2}^{\infty} a_k(f) \cdot \frac{S_k^{(q)}}{[n]_q}[F_{k-1}(z) - F_k(z)] \right| \leq \frac{C}{[n]_q^2}$$

holds, where $C > 0$ depends on f, r, G_r, and q but it is independent of n.

(iii) If $q \geq 1$, then for any $1 < r < \frac{R}{q}$, we have

$$\lim_{n \to \infty} [n]_q(\mathcal{B}_{n,q}(f;G)(z) - f(z)) = A_q(f)(z), \quad \text{uniformly in } \overline{G_r},$$

where $A_q(f)(z) = \sum_{k=2}^{\infty} a_k(f) \cdot S_k^{(q)} \cdot [F_{k-1}(z) - F_k(z)]$.

Proof. Let $1 < r < \frac{R}{q}$. Firstly, let us present the most important relationships which will be used in our proof.

By taking $R = r$ in the inequality (8), p. 43 in Suetin [136], we have

$$|F_p(z)| \leq C(r)r^p, \quad \text{for all } z \in \overline{G_r}, p \geq 0. \tag{1.11.2}$$

Indeed, by relationship (8) above mentioned (with r instead of R there), we have $|F_p(z)| \leq C(r)r^p$, for all $z \in \Gamma_r$, which, by the Maximum Modulus

Theorem for analytic functions, implies $|F_p(z)| \leq c(r)r^p$ for all $z \in \overline{G_r}$ (for these estimates, see also Curtiss [29], page 583, relationship (4.1) and the next two lines).

It is also worth noting that similar estimates hold from page 42, relationships (1), (3), and (4) in Suetin [136], by taking there $r = 1 + \varepsilon$ and $K = \overline{G_{r'}}$, with $1 < r' < r$ arbitrary close to r (in this case we get $|F_p(z)| \leq C(r)r^p$, for all $z \in \overline{G_{r'}}, p \geq 0$, but which still is good enough for the proof, because r in $1 < r < R$ is arbitrary and $r' < r$ is arbitrary close to r).

Let β satisfy $qr < \beta < R$. By (1.11.1), and by Cauchy's theorem, we obtain

$$|a_k(f)| = \left| \frac{1}{2\pi i} \int_{|u|=1} \frac{f(\Psi(u))}{u^{n+1}} du \right| = \left| \frac{1}{2\pi i} \int_{|u|=\beta} \frac{f(\Psi(u))}{u^{n+1}} du \right| \leq \frac{C(\beta, f)}{\beta^k},$$

(1.11.3)

for al $k \geq 0$, where $C(\beta, f) > 0$ is independent of k (see also the estimate in [49], p. 22).

Simple calculation shows that

$$S_k^{(q)} = \frac{k(k-1)}{2}, \text{ for } q = 1 \text{ and } S_k^{(q)} = \frac{q^k - k(q-1) - 1}{(q-1)^2}, \text{ for } q > 1.$$

(1.11.4)

Also, reasoning exactly as in the case of $q = 1$ (in [49], pp. 20–21), for all $q \geq 1$ we easily get

$$\mathcal{B}_{n,q}(f; G)(z) = \sum_{k=0}^{\infty} a_k(f) \mathcal{B}_{n,q}(F_k; G)(z).$$

(1.11.5)

Indeed, denoting $f_m(z) = \sum_{k=0}^{m} a_k(f) F_k(z)$, $m \in \mathbb{N}$, by the linearity of $\mathcal{B}_{n,q}$, we easily get

$$\mathcal{B}_{n,q}(f_m; G)(z) = \sum_{k=0}^{m} a_k(f) \mathcal{B}_{n,q}(F_k; G)(z), \text{ for all } z \in \overline{G_r}.$$

Therefore, it suffices to prove that $\lim_{m \to \infty} \mathcal{B}_{n,q}(f_m; G)(z) = \mathcal{B}_{n,q}(f; G)(z)$, for all $z \in \overline{G_r}$ and $n \in \mathbb{N}$.

Firstly, by the definition of $\mathcal{B}_{n,q}$, we have

$$\mathcal{B}_{n,q}(f_m; G)(z) = \sum_{p=0}^{n} \binom{n}{p}_q [\Delta_{1/[n]_q}^p G_m(0)]_q F_k(z),$$

where $G_m(w) = \frac{1}{2\pi i} \int_{|u|=1} \frac{f_m(\Psi(u))}{u - w} du$.

Note here that since by Gaier [38], p. 48, the first relation before (6.17), we have

$$\mathcal{F}_k(w) = \frac{1}{2\pi i} \int_{|u|=1} \frac{F_k(\Psi(u))}{u-w} du = w^k, \text{ for all } |w| < 1,$$

evidently that $\mathcal{F}_k(w)$ can be extended by continuity on $\partial \mathbb{D}_1$. This also immediately implies that $G_m(w) = \frac{1}{2\pi i} \int_{|u|=1} \frac{f_m(\Psi(u))}{u-w} du$ can be extended by continuity on $\partial \mathbb{D}_1$, which means that $\mathcal{B}_{n,q}(F_k; G)(z)$ and $\mathcal{B}_{n,q}(f_m; G)(z)$ are well defined.

Now, taking into account Cauchy's theorem we also can write

$$G_m(w) = \frac{1}{2\pi i} \int_{|u|=\beta} \frac{f_m(\Psi(u))}{u-w} du \text{ and } F(w) = \frac{1}{2\pi i} \int_{|u|=\beta} \frac{f(\Psi(u))}{u-w} du.$$

For all $n, m \in \mathbb{N}$ and $z \in \overline{G_r}$, it follows

$$|\mathcal{B}_{n,q}(f_m; G)(z) - \mathcal{B}_{n,q}(f; G)(z)| \leq \sum_{p=0}^{n} \binom{n}{p}_q |[\Delta_{1/[n]_q}^p (G_m - F)(0)]_q| \cdot |F_p(z)|$$

$$\leq \sum_{p=0}^{n} \binom{n}{p}_q \sum_{j=0}^{p} q^{j(j-1)/2} \binom{p}{j}_q |(G_m - F)([p-j]_q/[n]_q)| \cdot |F_p(z)|$$

$$\leq \sum_{p=0}^{n} \binom{n}{p}_q \sum_{j=0}^{p} \binom{p}{j}_q C_{j,q,\beta} \|f_m - f\|_{\overline{G_\beta}} \cdot |F_p(z)| \leq M_{n,q,r,\beta,G_\beta} \|f_m - f\|_{\overline{G_\beta}},$$

which, by $\lim_{m \to \infty} \|f_m - f\|_{\overline{G_\beta}} = 0$ (see, e.g., the proof of Theorem 2, p. 52 in Suetin [136]), implies (1.11.5) (here $\|f\|_{\overline{G_\beta}}$ denotes the uniform norm of f on $\overline{G_\beta}$).

Here, note that since by [49], p. 21 we have

$$\mathcal{F}_k(z) = \frac{1}{2\pi i} \int_{|u|=1} \frac{F_k(\Psi(u))}{u-z} du = e_k(z),$$

with $e_k(z) = z^k$, we easily get

$$\mathcal{B}_{n,q}(F_k; G)(z)$$

$$= \sum_{p=0}^{n} \binom{n}{p}_q [\Delta_{1/[n]_q}^p \mathcal{F}_k(0)]_q F_p(z) = \sum_{p=0}^{n} \binom{n}{p}_q [\Delta_{1/[n]_q}^p e_k(0)]_q F_p(z).$$

By the relationships (5), (6), and (7) in [118], p. 236, it immediately follows

$$\mathcal{B}_{n,q}(F_k; G)(z) = \sum_{p=0}^{n} D_{n,p,k}^{(q)} F_p(z),$$

where

$$D_{n,p,k}^{(q)} = \binom{n}{p}_q \cdot q^{p(p-1)/2} \cdot [0, 1/[n]_q, \ldots, p/[n]_q; e_k] \cdot \frac{[p]_q!}{[n]_q^p}$$

$$= \lambda_{n,p}^{(q)} \cdot [0, 1/[n]_q, \ldots, p/[n]_q; e_k], \qquad (1.11.6)$$

with $\lambda_{n,p}^{(q)} = \left(1 - \frac{[1]_q}{[n]_q}\right) \cdot \ldots \cdot \left(1 - \frac{[p-1]_q}{[n]_q}\right)$.

Note that by (1.11.6) (see also Lemma 3, p. 245 in [118]) we obtain

$$D_{n,p,k}^{(q)} \geq 0 \text{ for all } 0 \leq p \leq n, \ k \geq 0, \text{ and } \sum_{p=0}^{n} D_{n,p,k}^{(q)} = 1, \text{ for all } 0 \leq k \leq n$$

$$(1.11.7)$$

and

$$D_{n,k,k}^{(q)} = \Pi_{i=1}^{k-1}\left(1 - \frac{[i]_q}{[n]_q}\right), \ D_{n,k-1,k}^{(q)} = \frac{S_k^{(q)}}{[n]_q} \cdot \Pi_{i=1}^{k-2}\left(1 - \frac{[i]_q}{[n]_q}\right), \ k \leq n.$$

$$(1.11.8)$$

In what follows, first we prove that $A_q(f)(z)$ given by

$$A_q(f)(z) = \sum_{k=2}^{\infty} a_k(f) \cdot S_k^{(q)} \cdot [F_{k-1}(z) - F_k(z)],$$

is analytic in $\overline{G_r}$, for $1 < r < \frac{R}{q}$.

Indeed, by the inequality

$$|A_q(f)(z)| \leq \sum_{k=0}^{\infty} |a_k(f)| \cdot S_k^{(q)} \cdot [|F_{k-1}(z)| + |F_k(z)|],$$

since by (1.11.4) we get $S_k^{(q)} \leq \frac{q^k}{(q-1)^2}$ for $q > 1$, by (1.11.2) and (1.11.3), it immediately follows

$$|A_q(f)(z)| \leq \frac{2C(r) \cdot C(\beta, f)}{(q-1)^2} \cdot \sum_{k=0}^{\infty} d^k = \frac{2C(r) \cdot C(\beta, f)}{(1-d)(q-1)^2}, \text{ if } q > 1,$$

and

$$|A_q(f)(z)| \leq C(r) \cdot C(\beta, f) \sum_{k=0}^{\infty} k(k-1)d^k, \text{ if } q = 1,$$

with $d = \frac{rq}{\beta} < 1$, for all $z \in \overline{G_r}$. But by the ratio test the above series is uniformly convergent, which immediately shows that for $q \geq 1$, the function $A_q(f)$ is well defined and analytic in $\overline{G_r}$.

Now, by (1.11.5) we obtain

$$\left| \mathcal{B}_{n,q}(f;G)(z) - f(z) - \sum_{k=0}^{\infty} a_k(f) \cdot \frac{S_k^{(q)}}{[n]_q}[F_{k-1}(z) - F_k(z)] \right|$$

$$\leq \sum_{k=0}^{\infty} |a_k(f)| \cdot |E_{k,n}^{(q)}(G)(z)|,$$

where

$$E_{k,n}^{(q)}(G)(z) = \mathcal{B}_{n,q}(F_k;G)(z) - F_k(z) - \frac{S_k^{(q)}}{[n]_q}[F_{k-1}(z) - F_k(z)].$$

Because simple calculations implies that

$$E_{0,n}^{(q)}(G)(z) = E_{1,n}^{(q)}(G)(z) = E_{2,n}^{(q)}(G)(z) = 0,$$

in fact we have to estimate the expression

$$\sum_{k=3}^{\infty} |a_k(f)| \cdot |E_{k,n}^{(q)}(G)(z)|$$

$$= \sum_{k=3}^{n} |a_k(f)| \cdot |E_{k,n}^{(q)}(G)(z)| + \sum_{k=n+1}^{\infty} |a_k(f)| \cdot |E_{k,n}^{(q)}(G)(z)|.$$

To estimate $|E_{k,n}^{(q)}(G)(z)|$, we distinguish two cases: (i) $3 \leq k \leq n$; (ii) $k \geq n+1$.

Case 1. By using (1.11.2), we obtain

$$[n]_q|E_{k,n}^{(q)}(G)(z)| = |[n]_q(\mathcal{B}_{n,q}(F_k;G)(z) - F_k(z)) - S_k^{(q)} \cdot (F_{k-1} - F_k)|$$

$$\leq C(r)r^k[n]_q \sum_{i=1}^{k-2} D_{n,i,k}^{(q)} + C(r)r^k|[n]_q D_{n,k-1,k}^{(q)} - S_k^{(q)}|$$

$$+ C(r)r^k|[n]_q(1 - D_{n,k,k}^{(q)}) - S_k^{(q)}|.$$

Taking now into account (1.11.7) and (1.11.8) and following exactly the reasonings in the proof of Lemma 3, p. 747 in [145], we arrive at

$$|E_{k,n}^{(q)}(G)(z)| \leq \frac{4C(r)(k-1)^2[k-1]_q^2}{[n]_q^2} \cdot r^k, \text{ for all } z \in \overline{G_r}. \qquad (1.11.9)$$

Let β satisfy $qr < \beta < R$. By (1.11.3) and (1.11.9) it follows

$$\sum_{k=3}^{n} |a_k(f)| \cdot |E_{k,n}^{(q)}(G)(z)| \leq \frac{4C(r) \cdot C(\beta, f)}{[n]_q^2} \cdot \sum_{k=3}^{n}(k-1)^2[k-1]_q^2 \rho^k,$$

$$\qquad (1.11.10)$$

for all $z \in \overline{G_r}$ and $n \in \mathbb{N}$, where $\rho = \frac{r}{\beta} < \frac{1}{q}$.

Case 2. We get

$$\sum_{k=n+1}^{\infty} |a_k(f)| \cdot |E_{k,n}^{(q)}(G)(z)|$$

$$\leq \sum_{k=n+1}^{\infty} |a_k(f)| \cdot |\mathcal{B}_{n,q}(F_k; G)(z)| + \sum_{k=n+1}^{\infty} |a_k(f)| \cdot |F_k(z)|$$

$$+\frac{1}{[n]_q} \sum_{k=n+1}^{\infty} |a_k(f)| \cdot S_k^{(q)} \cdot |F_{k-1}(z)| + \frac{1}{[n]_q} \sum_{k=n+1}^{\infty} |a_k(f)| \cdot S_k^{(q)} \cdot |F_k(z)|$$

$$=: L_{1,q}(z) + L_{2,q}(z) + L_{3,q}(z) + L_{4,q}(z). \tag{1.11.11}$$

We have two subcases: (2_i) $q = 1$; (2_{ii}) $q > 1$.

Subcase (2_i). By (1.11.2), (1.11.3), and (1.11.7), for $1 < r < \beta < R$, denoting $\rho = \frac{r}{\beta}$, it immediately follows

$$L_{1,1}(z) \leq \sum_{k=n+1}^{\infty} |a_k(f)| \cdot \sum_{p=0}^{n} D_{n,p,k}^{(1)} |F_p(z)| \leq \frac{C(r, \beta, f)}{n^2} \sum_{k=n+1}^{\infty} (k-1)^2 \rho^k,$$

and similarly

$$L_{2,1}(z) \leq \frac{C(r, \beta, f)}{n^2} \sum_{k=n+1}^{\infty} (k-1)^2 \rho^k.$$

for all $z \in \overline{G_r}$.

Next, by similar reasonings as above and by (1.11.4), we obtain

$$L_{3,1}(z) \leq \frac{C(r, \beta, f)}{n} \sum_{k=n+1}^{\infty} \frac{k(k-1)}{2} \rho^k \leq \frac{C(r, \beta, f)}{n^2} \sum_{k=n+1}^{\infty} (k-1)^3 \rho^k,$$

and

$$L_{4,1}(z) \leq \frac{C(r, \beta, f)}{n^2} \sum_{k=n+1}^{\infty} (k-1)^3 \rho^k,$$

which by (1.11.11) implies that there exists a constant $K(r, \beta, f) > 0$ independent of n, such that

$$\sum_{k=n+1}^{\infty} |a_k(f)| \cdot |E_{k,n}^{(1)}(G)(z)| \leq \frac{K(r, \beta, f)}{n^2} \sum_{k=n+1}^{\infty} (k-1)^3 \rho^k,$$

for all $z \in \overline{G_r}$ and $n \in \mathbb{N}$.

But the sequence $\{a_n = \sum_{k=n+1}^{\infty} (k-1)^3 \rho^k, n \in \mathbb{N}\}$ is convergent to zero (therefore bounded by a constant $M > 0$ independent of n), as the remainder of the convergent series $\sum_{k=0}^{\infty} (k-1)^3 \rho^k$ (e.g., applying the ratio test), which will imply

$$\sum_{k=n+1}^{\infty} |a_k(f)| \cdot |E_{k,n}^{(1)}(G)(z)| \le \frac{M \cdot K(r,\beta,f)}{n^2},$$

for all $z \in \overline{G_r}$ and $n \in \mathbb{N}$.

Now, taking $q = 1$ in (1.11.10) and taking int account that by the ratio test the series $\sum_{k=3}^{\infty}(k-1)^4\rho^k$ is convergent, we get

$$\sum_{k=3}^{n} |a_k(f)| \cdot |E_{k,n}^{(1)}(G)(z)| \le \frac{4C(r) \cdot C(\beta,f)}{n^2} \cdot \sum_{k=3}^{n}(k-1)^4\rho^k$$

$$\le \frac{4C(r) \cdot C(\beta,f)}{n^2} \cdot \sum_{k=3}^{\infty}(k-1)^4\rho^k = \frac{K'(r,\beta,f)}{n^2},$$

which combined with the previous estimate immediately implies

$$\sum_{k=0}^{\infty} |a_k(f)| \cdot |E_{k,n}^{(1)}(G)(z)| \le \frac{C}{n^2},$$

for all $z \in \overline{G_r}$ and $n \in \mathbb{N}$, where $C > 0$ is a constant independent of n.

This proves (i) in the statement of Theorem 1.11.4.

Subcase (2_{ii}). By (1.11.2), (1.11.3), and (1.11.7), for $1 < r < \frac{\beta}{q} < \frac{R}{q}$ and denoting $\rho = \frac{r}{\beta} < \frac{1}{q} < 1$, for all $z \in \overline{G_r}$, it follows

$$L_{1,q}(z) \le \sum_{k=n+1}^{\infty} |a_k(f)| \cdot \sum_{p=0}^{n} D_{n,p,k}^{(q)}|F_p(z)| \le C(r,\beta,f) \sum_{k=n+1}^{\infty} \rho^k$$

$$\le \frac{C(r,\beta,f)}{[n]_q^2} \sum_{k=n+1}^{\infty} [k-1]_q^2\rho^k \le \frac{C(r,\beta,f,q)}{[n]_q^2} \sum_{k=n+1}^{\infty} (q^2\rho)^k$$

and similarly

$$L_{2,q}(z) \le C(r,\beta,f) \sum_{k=n+1}^{\infty} \rho^k \le \frac{C(r,\beta,f)}{[n]_q^2} \sum_{k=n+1}^{\infty} [k-1]_q^2\rho^k$$

$$\le \frac{C(r,\beta,f,q)}{[n]_q^2} \sum_{k=n+1}^{\infty} (q^2\rho)^k.$$

Also, since by (1.11.4) we get $S_k^{(q)} \le \frac{q^k}{(q-1)^2}$, by using (1.11.2) and (1.11.3) too, for all $z \in \overline{G_r}$, it follows

$$L_{3,q}(z) \leq \frac{C(r,\beta,f,q)}{[n]_q} \sum_{k=n+1}^{\infty} q^k \rho^k = \frac{C(r,\beta,f,q)}{[n]_q} \sum_{k=n+1}^{\infty} (q\rho)^k$$

$$\leq \frac{C'(r,\beta,f,q)}{[n]_q^2} \sum_{k=n+1}^{\infty} (q^2\rho)^k,$$

and

$$L_{4,q}(z) \leq \frac{C(r,\beta,f,q)}{[n]_q} \sum_{k=n+1}^{\infty} q^k \rho^k = \frac{C(\beta,f,q)}{[n]_q} \sum_{k=n+1}^{\infty} (q\rho)^k$$

$$\leq \frac{C'(r,\beta,f,q)}{[n]_q^2} \sum_{k=n+1}^{\infty} (q^2\rho)^k.$$

By (1.11.11), we immediately obtain

$$\sum_{k=n+1}^{\infty} |a_k(f)| \cdot |E_{k,n}^{(q)}(G)(z)| \leq \frac{K(r,\beta,f,q)}{[n]_q^2},$$

for all $z \in \overline{G_r}$, if $q^2\rho < 1$ (which holds for $1 < r < \frac{\beta}{q^2} < \frac{R}{q^2}$). Also, by (1.11.10), since $[k-1]_q^2 \leq [k]_q^2 \leq \frac{q^{2k}}{(q-1)^2}$, for $z \in \overline{G_r}$ with $1 < r < \frac{\beta}{q^2} < \frac{R}{q^2}$, we easily obtain

$$\sum_{k=3}^{n} |a_k(f)| \cdot |E_{k,n}^{(q)}(G)(z)| \leq \frac{K'(r,\beta,f,q)}{[n]_q^2}.$$

Collecting these results, we immediately obtain the upper estimate in (ii).

(iii) The case $q = 1$ follows directly by multiplying by n in the estimate in (i) and by passing to limit with $n \to \infty$. In the case of $q > 1$, if $1 < r < \frac{R}{q^2}$, then by multiplying in (ii) with $[n]_q$ and passing to limit with $n \to \infty$, we get the desired conclusion.

What remained to be proved is that the limit in (iii) still holds under the more general condition $1 < r < \frac{R}{q}$.

Since $\frac{R}{q^{1+t}} \nearrow \frac{R}{q}$ as $t \searrow 0$, evidently that given $1 < r < \frac{R}{q}$, there exists a $t \in (0,1)$, such that $q^{1+t}r < R$. Because f is analytic in G, choosing β with $q^{1+t}r < \beta < R$, by (1.11.2) and (1.11.3), this implies that $\sum_{k=2}^{\infty} |a_k(f)| k^4 q^{(1+t)k} r^k \leq \sum_{k=2}^{\infty} k^4 \left(\frac{q^{1+t}r}{\beta}\right)^k < \infty$, for all $z \in \overline{G_r}$. Also, the convergence of the previous series implies that for arbitrary $\varepsilon > 0$, there exists n_0, such that $\sum_{k=n_0+1}^{\infty} |a_k(f)| k^2 q^k r^k < \varepsilon$.

By using (1.11.9), for all $z \in \overline{G_r}$ and $n > n_0$ we get

$$|[n]_q (\mathcal{B}_{n,q}(f;G)(z) - f(z)) - A_q(f)(z)|$$

$$\leq \sum_{k=2}^{n_0} |a_k(f)| \cdot \left| [n]_q (\mathcal{B}_{n,q}(F_k;G)(z) - F_k(z)) - S_k^{(q)}[F_{k-1}(z) - F_k(z)] \right|$$

$$+ \sum_{k=n_0+1}^{\infty} |a_k(f)| \cdot \left([n]_q |\mathcal{B}_{n,q}(F_k; G)(z) - F_k(z)| + S_k^{(q)} |F_{k-1}(z) - F_k(z)| \right)$$

$$\leq C(r) \sum_{k=2}^{n_0} |a_k(f)| \cdot \frac{4(k-1)^2[k-1]_q^2}{[n]_q} \cdot r^k$$

$$+ \sum_{k=n_0+1}^{\infty} |a_k(f)| \cdot \left([n]_q |\mathcal{B}_{n,q}(F_k; G)(z) - F_k(z)| + S_k^{(q)} |F_{k-1}(z) - F_k(z)| \right).$$

But by the proof of Theorem 1.11.4, Case 1, with $\alpha = \gamma = 0$, in [61], for $k \leq n$, we have

$$|\mathcal{B}_{n,q}(F_k; G)(z) - F_k(z)| \leq C(r) \frac{k[k-1]_q}{[n]_q} \cdot r^k,$$

while for $k > n$ and using (1.11.2) and (1.11.7), we get

$$|\mathcal{B}_{n,q}(F_k; G)(z) - F_k(z)|$$

$$\leq |\mathcal{B}_{n,q}(F_k; G)(z)| + |F_k(z)| \leq \sum_{p=0}^{n} D_{n,p,k}^{(q)} |F_p(z)| + |F_k(z)|$$

$$\leq C(r)r^n + C(r)r^k \leq C'(r)r^k \leq 2C'(r) \frac{k[k-1]_q}{[n]_q} \cdot r^k,$$

for all $z \in \overline{G_r}$.

Also, since $S_k^{(q)} \leq (k-1)[k-1]_q$, by using (1.11.2), it is immediate that

$$S_k^{(q)} \cdot |F_{k-1}(z) - F_k(z)| \leq S_k^{(q)} \cdot [|F_{k-1}(z)| + |F_k(z)|]$$

$$\leq 2C(r)(k-1)[k-1]_q r^k.$$

Therefore, we easily obtain

$$\sum_{k=n_0+1}^{\infty} |a_k(f)| \cdot \left([n]_q |\mathcal{B}_{n,q}(F_k; G)(z) - F_k(z)| + S_k^{(q)} |F_{k-1}(z) - F_k(z)| \right)$$

$$\leq K(r) \cdot \sum_{k=n_0+1}^{\infty} |a_k(f)| \cdot (k-1)[k-1]_q r^k,$$

valid for all $z \in \overline{G_r}$, where $K(r) > 0$ is a constant depending only on r.

In conclusion, for all $z \in \overline{G_r}$ and $n > n_0$ we have

$$|[n]_q(\mathcal{B}_{n,q}(f; G)(z) - f(z)) - A_q(f)(z)|$$

$$\leq C(r) \sum_{k=2}^{n_0} |a_k(f)| \cdot \frac{4(k-1)^2[k-1]_q^2}{[n]_q} \cdot r^k$$

$$+K(r) \cdot \sum_{k=n_0+1}^{\infty} |a_k(f)| \cdot (k-1)[k-1]_q r^k$$

$$\leq \frac{4C(r)}{[n]_q^t} \cdot \sum_{k=2}^{n_0} |a_k(f)| \cdot k^2 [k-1]_q^{1+t} \cdot r^k + K(r) \cdot \sum_{k=n_0+1}^{\infty} |a_k(f)| \cdot k^2 q^k r^k$$

$$\leq \frac{4C(r)}{[n]_q^t} \cdot \sum_{k=2}^{\infty} |a_k(f)| \cdot k^4 q^{(1+t)k} \cdot r^k + K(r)\varepsilon.$$

Now, since $\frac{4C(r)}{[n]_q^t} \to 0$ as $n \to \infty$ and $\sum_{k=2}^{\infty} |a_k(f)| \cdot k^4 q^{(1+t)k} \cdot r^k < \infty$, for the given $\varepsilon > 0$, there exists an index n_1, such that

$$\frac{4C(r)}{[n]_q^t} \cdot \sum_{k=2}^{\infty} |a_k(f)| \cdot k^4 q^{(1+t)k} \cdot r^k < \varepsilon,$$

for all $n > n_1$.

As a final conclusion, for all $n > \max\{n_0, n_1\}$ and $z \in \overline{G_r}$, we get

$$\left| [n]_q (\mathcal{B}_{n,q}(f; G)(z) - f(z)) - A_q(f)(z) \right| \leq (1 + K(r))\varepsilon,$$

which shows that

$$\lim_{n \to \infty} [n]_q (\mathcal{B}_{n,q}(f; G)(z) - f(z)) = A_q(f)(z), \text{ uniformly in } \overline{G_r}.$$

The theorem is proved. □

As a consequence of Theorem 1.11.4, we get the exact order of approximation, by the following.

Corollary 1.11.5. *Under the hypothesis of Theorem 1.11.4, suppose that $R > q \geq 1$. If $1 < r < \frac{R}{q}$ and f is not a polynomial of degree ≤ 1 in G, then*

$$\|\mathcal{B}_n(f; G) - f\|_{\overline{G_r}} \sim \frac{1}{[n]_q}, \quad n \in \mathbb{N},$$

holds, where $\|f\|_{\overline{G_r}} = \sup\{|f(z)|; z \in \overline{G_r}\}$ and the constants in the equivalence depend on f, r, and G_r but are independent of n.

Proof. Suppose that we would have $\|\mathcal{B}_n(f; G) - f\|_{\overline{G_r}} = o([n]_q^{-1})$. Then, combining Theorem 1.11.4, (iii) with Theorem 1.11.2 in the case $q = 1$ and with Theorem 1.11.3 in the case $q > 1$, would immediately follow that $A_q(f) = 0$ for all $z \in \overline{G_r}$, where $A_q(f)(z)$ is defined in the statement of Theorem 1.11.4, (iii).

But $A_q(f)(z) = 0$ for all $z \in \overline{G_r}$ by simple calculation implies

$$2a_2(f) S_2^{(q)} F_1(z) + \sum_{k=2}^{\infty} [S_{k+1}^{(q)} a_{k+1}(f) - S_k^{(q)} a_k(f)] F_k(z) = 0, \quad z \in \overline{G_r}.$$

By the uniqueness of Faber polynomial series (see [34], or [136], pp. 108–109), since by (1.11.4) it is clear that $S_k^{(q)} > 0$ for all $k \geq 2$, we would get that $a_2(f) = 0$ and

$$S_{k+1}^{(q)} a_{k+1}(f) - S_k^{(q)} a_k(f) = 0, \text{ for all } k = 2, 3, \ldots,.$$

For $k = 2$ we easily get $a_3(f) = 0$, and taking above step by step $k = 3, 4, \ldots,$ we easily would obtain that $a_k(f) = 0$ for all $k \geq 2$.

Therefore we would get $f(z) = a_0(f)F_0(z) + a_1(f)F_1(z)$ for all $z \in \overline{G_r}$. But because $F_k(z)$ is a polynomial of exact degree k, it would follow that f would be a polynomial of degree ≤ 1 in $\overline{G_r}$, a contradiction with the hypothesis. In conclusion, if f is not a polynomial of degree ≤ 1, then by the Theorems 1.11.2 and 1.11.3, the approximation order is exactly $\frac{1}{[n]_q}$, which proves the corollary.   □

Remarks. 1) The q-Bernstein–Faber polynomials with $q \geq 1$ in Definition 1.11.1 together with Theorem 1.11.4 represent natural extensions to compact subsets in $\mathbb{C}$ of the classical Bernstein polynomials and of Voronovskaja's formula on $[0, 1]$, respectively.

2) By the obvious inequalities $\frac{q-1}{q^n} \leq \frac{1}{[n]_q} \leq \frac{q}{q^n}$, for all $n \in \mathbb{N}$ and $q > 1$, it follows that if f is not a polynomial of degree ≤ 1, then $\mathcal{B}_{n,q}(f; G)$ approximates f in the continuum G, with the exact order $\frac{1}{q^n}$.

3) For $G = \overline{\mathbb{D}}_r$, $r > 1$, Corollary 1.11.5 becomes Corollary 1 in [145].

4) Clearly the above Theorems 1.11.2, 1.11.3, and 1.11.4 and Corollary 1.11.5 can be applied to the concrete examples already mentioned in Sect. 1.9 by Applications 1.9.6–1.9.9.

As a general conclusion, based on the previous concrete examples too, we can say that in approximation theory, the q-Bernstein–Faber polynomials, $q > 1$, attached to a function defined on a subset of the complex plane and given by Definition 1.11.1, can represent a good alternative to the partial sums of the Faber series attached to the same function and subset, both giving a geometric progression order of approximation.

1.12 q-Bernstein Polynomials of Quaternion Variable, $q \geq 1$

As it is well known, the field of complex numbers can be extended to more general algebraic structures (with several complex units) called hypercomplex numbers. These structures can be divided in two main classes: commutative hypercomplex structures which are rings with divisors of zeros and noncommutative hypercomplex structures which are fields (so without divisors of zero), the most known being the so-called quaternion numbers and Clifford algebras.

Therefore, it is natural to see for extensions of the approximation results in the previous sections, to approximation by q-Bernstein-type operators of hypercomplex variables, $q \geq 1$. In this section we limit our consideration to the case of q-Bernstein polynomials of quaternion variable, $q > 1$, because the case $q = 1$ was considered in Gal [49], Chapter 4, Section 4.1. For this purpose first we make a short introduction.

The quaternion field is defined by

$$\mathbb{H} = \{z = x_1 + x_2 i + x_3 j + x_4 k; x_1, x_2, x_3, x_4 \in \mathbb{R}\},$$

where the complex units $i, j, k \notin \mathbb{R}$ satisfy

$$i^2 = j^2 = k^2 = -1, \ ij = -ji = k, \ jk = -kj = i, \ ki = -ik = j.$$

It is a noncommutative field, and since obviously $\mathbb{C} \subset \mathbb{H}$, it extends the class of complex numbers. On $\mathbb{H}$ can be defined the norm $\|z\| = \sqrt{x_1^2 + x_2^2 + x_3^3 + x_4^2}$, for $z = x_1 + x_2 i + x_3 j + x_4 k$.

If $G \subset \mathbb{H}$, then a function $f : G \to \mathbb{H}$ can be written in the form

$$f(z) =$$

$$f_1(x_1, x_2, x_3, x_4) + f_2(x_1, x_2, x_3, x_4)i + f_3(x_1, x_2, x_3, x_4)j + f_4(x_1, x_2, x_3, x_4)k,$$

$z = x_1 + x_2 i + x_3 j + x_4 k \in G$, where f_i are real valued functions, $i = 1, 2, 3, 4$.

The fact that a direct attempt to generalize the concept of differentiability for f as

$$\lim_{z \to z_0} (z - z_0)^{-1}[f(z) - f(z_0)] \in \mathbb{H}, z_0 \in G,$$

or as

$$\lim_{z \to z_0} [f(z) - f(z_0)](z - z_0)^{-1} \in \mathbb{H}, z_0 \in G,$$

necessarily implies that f is of the form $f(z) = Az + B$ (see Mejlihzon [108]) is well known. For this reason, the theory of holomorphic functions of quaternion variable can be constructed in several other ways, by producing different classes of holomorphic functions. We mention below only two ways. The first one is given by the following.

Definition 1.12.1 (Moisil [112]). Let $f = f_1 + f_2 i + f_3 j + f_4 k$ be such that each f_i has continuous partial derivatives of order one, $i = 1, 2, 3, 4$. Define $F = \frac{\partial}{\partial x_1} + \frac{\partial}{\partial x_2}i + \frac{\partial}{\partial x_3}j + \frac{\partial}{\partial x_4}k$. One says that f is left differentiable (monogenic) at z_0 if $Ff(z_0) = 0$. In this case, the derivative of f at z_0 will be given $f'(z_0) = \overline{D}(f)(z_0)$, where $\overline{F} = \frac{\partial}{\partial x_1} - \frac{\partial}{\partial x_2}i - \frac{\partial}{\partial x_3}j - \frac{\partial}{\partial x_4}k$.

If f is monogenic at each z, then it is called holomorphic.

Remarks. 1) In the case of complex variable, in the differential operator F, one reduces to the areolar derivative of f, and the operator $\overline{F}$ becomes the derivative of f.

2) f will be called right differentiable at z_0 if $fF(z_0) = 0$.

The second kind of definition for holomorphy is suggested by Weierstrass's idea in the case of complex variable.

Definition 1.12.2. Denoting $\mathbb{D}_R = \{z \in \mathbb{H}; \|z\| < R\}$, one says that $f : \mathbb{D}_R \to \mathbb{H}$ is left W-analytic in $\mathbb{D}_R$ if $f(z) = \sum_{k=0}^{\infty} c_k z^k$, for all $z \in \mathbb{D}_R$, where $c_k \in \mathbb{H}$ for all $k = 0, 1, 2, \ldots,$. Also, f is called right W-analytic in $\mathbb{D}_R$ if $f(z) = \sum_{k=0}^{\infty} z^k c_k$, for all $z \in \mathbb{D}_R$.

Here the convergence of the partial sums $\sum_{k=0}^{n} c_k z^k$ and $\sum_{k=0}^{n} z^k c_k$ to f is understood uniformly in any closed ball $\overline{\mathbb{D}_r} = \{z \in \mathbb{H}; \|z\| \leq r\}, 0 < r < R$, with respect to the metric $d(x, y) = \|x - y\|$.

If $f(z) = \sum_{k=0}^{n} c_k z^k$ $(f(z) = \sum_{k=0}^{n} z^k c_k)$, then f is called left (respectively right) polynomial of degree $\leq n$.

Remark. Note that according to Theorem 2.7 in Gentili and Struppa [70], a function of quaternionic variable is right W-analytic if and only if it is regular in the sense of Definition 1.1 in [70]. For details concerning the properties of the right W-analytic functions, see, e.g., Gentili and Stoppato [69].

Remark. While in the case of complex variable, the two concepts in Definitions 1.12.1. and 1.12.2 coincide, in the case of quaternion variable, this fact does not happen. The most suitable concept for our purpose is that in Definition 1.12.2.

Recall here the basic things in q-analysis we need. Let everywhere in this section $q \geq 1$. For any $n = 1, 2, \ldots$, define the q-integer $[n]_q := 1 + q + \ldots + q^{n-1}$, $[0]_q := 0$ and the q-factorial $[n]_q! := [1]_q [2]_q \ldots [n]_q$, $[0]_q! := 1$. For $q = 1$, we obviously get $[n]_q = n$.

For integers $0 \leq k \leq n$, define

$$\binom{n}{k}_q := \frac{[n]_q!}{[k]_q! [n-k]_q!}.$$

Evidently, for $q = 1$, we get $[n]_1 = n$, $[n]_1! = n!$ and $\binom{n}{k}_1 = \binom{n}{k}$.

Concerning the q-Bernstein polynomials, $q \geq 1$, due to noncommutativity, for $R > 1$ to a function $f : \mathbb{D}_R \to \mathbb{H}$, three distinct q-Bernstein polynomials can be attached, as follows:

$$B_{n,q}(f)(z) = \sum_{l=0}^{n} f\left(\frac{[l]_q}{[n]_q}\right) \binom{n}{l}_q z^l \Pi_{s=0}^{n-1-l} (1 - q^s z), z \in \mathbb{H},$$

$$B_{n,q}^*(f)(z) = \sum_{l=0}^{n} \binom{n}{l}_q z^l \Pi_{s=0}^{n-1-l}(1 - q^s z) f\left(\frac{[l]_q}{[n]_q}\right), z \in \mathbb{H},$$

$$B_{n,q}^{**}(f)(z) = \sum_{l=0}^{n} \binom{n}{l}_q z^l f\left(\frac{[l]_q}{[n]_q}\right) \Pi_{s=0}^{n-1-l}(1 - q^s z), z \in \mathbb{H}.$$

We may call them as the left q-Bernstein polynomials, right q-Bernstein polynomials, and middle q-Bernstein polynomials, respectively.

It is easy to show by a simple example that these kinds of q-Bernstein polynomials do not converge for any continuous function f. Indeed, if, for example, we take $q = 1$ and $f(z) = izi$, then we easily get

$$|B_{n,q}(f)(z) - izi| = |B_{n,q}^*(f)(z) - izi| = |B_{n,q}^{**}(f)(z) - izi| = |-z - izi| =$$

$$|-iz + zi| > 0, \quad \text{for all } z \neq i.$$

However, for each kind of q-Bernstein polynomial, there exists a suitable class of functions for which the convergence holds. To prove that, we need some auxiliary results.

Theorem 1.12.3 (Gal [60]). *Let $q \geq 1$. Suppose that $f : \mathbb{D}_R \to \mathbb{H}$ has the property that $f(z) \in \mathbb{R}$ for all $z \in [0,1]$. Then we have the representation formula*

$$B_{n,q}(f)(z) = \sum_{m=0}^{n} \binom{n}{m}_q [\Delta_{1/[n]_q}^m f(0)]_q z^m, \text{ for all } z \in \mathbb{H},$$

where $[\Delta_h^p f(0)]_q = \sum_{k=0}^{p}(-1)^k q^{k(k-1)/2} \binom{p}{k}_q f([p-k]_q h)$.

Proof. Because of the hypothesis on f, the real values $f\left(\frac{l}{[n]_q}\right)$ commute with the other terms in the expression of $B_n(f)(z)$, so that taking into account that $z^{n+m} = z^n z^m = z^m z^n$ (from associativity), $z^l \Pi_{s=0}^{n-1-l}(1 - q^s z) = \Pi_{s=0}^{n-1-l}(1 - q^s z)z^l$, $\alpha z = z\alpha$, for all $\alpha \in \mathbb{R}$, $z \in \mathbb{H}$ and that in the product $\Pi_{s=0}^{n-1-l}(1 - q^s z)$, we can interchange the order of the terms, reasoning exactly as in the case of q-Bernstein polynomials of real variable (see Phillips [123], proof of Theorem 1), we obtain that the coefficient of z^m in the expression of $B_{n,q}(f)(z)$ is

$$\sum_{r=0}^{m} f([m-r]_q/[n]_q) \binom{n}{m-r}_q (-1)^r q^{r(r-1)/2} \binom{n-m+r}{r}_q$$

$$= \binom{n}{m}_q \sum_{r=0}^{m}(-1)^r q^{r(r-1)/2} \binom{m}{r}_q f([m-r]_q/[n]_q),$$

which immediately proves the theorem. $\qquad\square$

Remark. It is clear that Theorem 1.12.3 holds for the right and middle q-Bernstein polynomials, $B_{n,q}^{*}(f)(z)$ and $B_{n,q}^{**}(f)(z)$, too.

Now we are in position to state the following approximation result.

Theorem 1.12.4 (Gal [60]). *Let $1 < q < R$ and suppose that $f : \mathbb{D}_R \to \mathbb{H}$ is left W-analytic in $\mathbb{D}_R$, i.e., $f(z) = \sum_{k=0}^{\infty} a_k z^k$, for all $z \in \mathbb{D}_R$, where $a_k \in \mathbb{H}$ for all $k = 0, 1, 2, \ldots,$. Then for all $1 < r < \frac{R}{q}$, $\|z\| \le r$ and $n \in \mathbb{N}$, we have*

$$\|B_{n,q}(f)(z) - f(z)\| \le \frac{2}{(q-1)[n]_q} \cdot \sum_{k=1}^{\infty} \|a_k\| k (qr)^k.$$

Proof. Denoting $e_k = z^k$, firstly we will prove that

$$B_{n,q}(f)(z) = \sum_{k=0}^{\infty} a_k B_{n,q}(e_k)(z), \text{ for all } \|z\| \le r.$$

In this sense denote by $f_m(z) = \sum_{k=0}^{m} a_k e_k(z)$, $m \in \mathbb{N}$, the partial sum of the Taylor expansion of f.

Since by the linearity of $B_{n,q}$, we easily get

$$B_{n,q}(f_m)(z) = \sum_{k=0}^{m} a_k B_{n,q}(e_k)(z), \text{ for all } \|z\| \le r,$$

it suffices to prove that $\lim_{m \to \infty} B_{n,q}(f_m)(z) = B_{n,q}(f)(z)$, for all $\|z\| \le r$ and $n \in \mathbb{N}$.

Firstly, by Theorem 1.12.3 we have

$$B_{n,q}(f_m)(z) = \sum_{p=0}^{n} \binom{n}{p}_q [\Delta_{1/[n]_q}^{p} f_m(0)]_q e_p(z).$$

For all $n, m \in \mathbb{N}$ and $\|z\| \le R$, it follows

$$\|B_{n,q}(f_m)(z) - B_{n,q}(f)(z)\| \le \sum_{p=0}^{n} \binom{n}{p}_q \|[\Delta_{1/[n]_q}^{p} (f_m - f)(0)]_q\| \cdot \|e_p(z)\|$$

$$\le \sum_{p=0}^{n} \binom{n}{p}_q \sum_{j=0}^{p} q^{j(j-1)/2} \binom{p}{j}_q \|(f_m - f)([p-j]_q/[n]_q)\| \cdot \|e_p(z)\| \le$$

$$\sum_{p=0}^{n} \binom{n}{p}_q \sum_{j=0}^{p} q^{j(j-1)/2} \binom{p}{j}_q C_{j,p,\beta} \||f_m - f\||_r \cdot \|e_p(z)\| \le M_{n,p,r,q} \||f_m - f\||_r,$$

which by $\lim_{m \to \infty} \||f_m - f\||_r = 0$ implies the desired conclusion. Here $\||f_m - f\||_r = \max\{\|f_m(z) - f(z)\|; \|z\| \le r\}$.

Consequently we obtain

$$\|B_{n,q}(f)(z) - f(z)\| \leq \sum_{k=0}^{\infty} \|a_k\| \cdot \|B_{n,q}(e_k)(z) - e_k(z)\|$$

$$= \sum_{k=0}^{n} \|a_k\| \cdot \|B_{n,q}(e_k)(z) - e_k(z)\| + \sum_{k=n+1}^{\infty} \|a_k\| \cdot \|B_{n,q}(e_k)(z) - e_k(z)\|.$$

Therefore it remains to estimate $\|B_{n,q}(e_k)(z) - e_k(z)\|$, firstly for all $0 \leq k \leq n$ and secondly for $k \geq n+1$, where

$$B_{n,q}(e_k)(z) = \sum_{p=0}^{n} \binom{n}{p}_q [\Delta^p_{1/[n]_q} e_k(0)]_q \cdot e_p(z).$$

Denote

$$D^{(q)}_{n,p,k} = \binom{n}{p}_q [\Delta^p_{1/[n]_q} e_k(0)]_q.$$

By relationship (12), p. 513 in Phillips [123], we can write

$$D^{(q)}_{n,p,k} = q^{p(p-1)/2} \binom{n}{p}_q [0, [1]_q/[n]_q, \ldots, [p]_q/[n]_q; e_k] \cdot ([p]_q!)/[n]_q^p$$

$$= \left(1 - \frac{[1]_q}{[n]_q}\right) \cdots \left(1 - \frac{[p-1]_q}{[n]_q}\right) [0, [1]_q/[n]_q, \ldots, [p]_q/[n]_q; e_k],$$

where $[0, [1]_q/[n]_q, \ldots, [p]_q/[n]_q; e_k]$ denotes the divided difference of $e_k(z) = z^k$.

It follows

$$B_{n,q}(e_k)(z) = \sum_{p=0}^{n} D^{(q)}_{n,p,k} \cdot e_p(z).$$

Since by (6) and (7) p. 236 in Ostrovska [117], for the classical complex q-Bernstein polynomials, $B_{n,q}(f)(z)$, attached to a disk of center in origin, we also can write $B_{n,q}(e_k)(z) = \sum_{p=0}^{n} D^{(q)}_{n,p,k} z^p$, since each e_k is convex of any order and $B_{n,q}(e_k)(1) = e_k(1) = 1$ for all k, it follows that all $D^{(q)}_{n,p,k} \geq 0$ and $\sum_{p=0}^{n} D^{(q)}_{n,p,k} = 1$, for all k and n.

Also, note that $D^{(q)}_{n,k,k} = \left(1 - \frac{[1]_q}{[n]_q}\right) \cdots \left(1 - \frac{[k-1]_q}{[n]_q}\right)$, for all $k \geq 1$ and that $D^{(q)}_{n,0,0} = 1$.

In the estimation of $\|B_{n,q}(e_k)(z) - e_k(z)\|$, we distinguish two cases: 1) $0 \leq k \leq n$; 2) $k > n$.

Case 1. We have

$$\|B_{n,q}(e_k)(z) - e_k(z)\| \leq \|e_k(z)\| \cdot |1 - D_{n,k,k}^{(q)}| + \sum_{p=0}^{k-1} D_{n,p,k}^{(q)} \cdot \|e_p(z)\|.$$

Since $\|e_p(z)\| \leq r^p$, for all $\|z\| \leq r$ and $p \geq 0$, by the relationships in the proof of Theorem 5, p. 247 and by Corollary 6, p. 244, both in Ostrovska [117], we immediately get

$$\|B_{n,q}(e_k)(z) - e_k(z)\| \leq 2[1 - D_{n,k,k}^{(q)}]r^k \leq 2\frac{(k-1)[k-1]_q}{[n]_q}r^k$$

$$\leq 2\frac{kq^k}{(q-1)[n]_q}r^k = 2\frac{k(qr)^k}{(q-1)[n]_q},$$

for all $\|z\| \leq r$.
Case 2. We have

$$\|B_{n,q}(e_k)(z) - e_k(z)\| \leq \|B_{n,q}(e_k)(z)\| + \|e_k(z)\| \leq 2r^k$$

$$\leq 2\frac{(k-1)[k-1]_q}{[n]_q}r^k \leq 2\frac{(k-1)[k-1]_q}{[n]_q}(qr)^k \leq 2\frac{k(qr)^k}{(q-1)[n]_q}.$$

In conclusion, collecting the estimates in Cases 1 and 2, we obtain

$$\|B_{n,q}(f)(z) - f(z)\| \leq \frac{2}{(q-1)[n]_q} \cdot \sum_{k=0}^{\infty} \|a_k\|k(qr)^k, \|z\| \leq r, n \in \mathbb{N},$$

which proves the theorem. □

In a similar manner we obtain the following.

Corollary 1.12.5 (Gal [60]). *Let* $1 < q < R$ *and suppose that* $f : \mathbb{D}_R \to \mathbb{H}$ *is right W-analytic in* $\mathbb{D}_R$, *i.e.,* $f(z) = \sum_{p=0}^{\infty} z^p a_p$, *for all* $z \in \mathbb{D}_R$, *where* $c_p \in \mathbb{H}$ *for all* $p = 0, 1, 2, \ldots,$. *Then for all* $1 \leq r < \frac{R}{q}$, $\|z\| \leq r$ *and* $n \in \mathbb{N}$, *we have*

$$\|B_{n,q}^*(f)(z) - f(z)\| \leq \frac{2}{(q-1)[n]_q} \cdot \sum_{k=1}^{\infty} \|a_k\|k(qr)^k.$$

Remarks. 1) Taking into account the inequality $\frac{1}{[n]_q} \leq q \cdot \frac{1}{q^n}$ for all $n \in \mathbb{N}$, it follows that the order of approximation in Theorem 1.12.4 and in Corollary 1.12.5 is $\frac{1}{q^n}$.

2) It is not difficult to see that in the case of Bernstein-type polynomials $B_n^{**}(f)(z)$, an estimate of the form in Theorem 1.12.4 cannot be obtained, because in general we cannot write a formula of the type

$B_n^{**}(f)(z) = \sum_{p=0}^{\infty} c_p B_n^{**}(e_p)(z)$ for f left W-analytic or of the type
$B_n^{**}(f)(z) = \sum_{p=0}^{\infty} B_n^{**}(e_p)(z)c_p$ for f right W-analytic.

For the proof of our next results, we also need the following known result.

Theorem 1.12.6 (Gal [49], p. 297–298, Theorem 4.1.4). *Suppose that* $f : \mathbb{D}_R \to \mathbb{H}$ *is left W-analytic in* $\mathbb{D}_R$, *i.e.,* $f(z) = \sum_{p=0}^{\infty} c_p z^p$, *for all* $z \in \mathbb{D}_R$, *where* $c_p \in \mathbb{H}$ *for all* $p = 0, 1, 2, \ldots,$. *Then for all* $1 \leq r < R$, $\|z\| \leq r$ *and* $n \in \mathbb{N}$, *we have*

$$\|B_{n,1}(f)(z) - f(z)\| \leq \frac{2}{n} \sum_{p=2}^{\infty} \|c_p\| p(p-1) r^p = O(1/n).$$

Remark. If f is supposed right W-analytic, then a similar upper estimate in approximation by the right $B_{n,1}^*(f)(z)$ operators was obtained by Gal [49], p. 299, Corollary 4.1.5.

Now, in what follows, we continue our investigations by obtaining Voronovskaja-type results for $B_{n,q}(f)(z)$ and $B_{n,q}^*(f)(z)$, $q \geq 1$. Then, as applications will follow that for functions which are not left polynomials of degree ≤ 1, the orders of approximation in Theorems 1.12.6 and 1.12.4 are exactly $\frac{1}{n}$ and $\frac{1}{q^n}$, respectively. The Voronovskaja-type results obtained below are extensions of those for the q-Bernstein operators ($q \geq 1$) of complex variable in Gal [41] and Wang and Wu [145], to the case of a quaternionic variable.

Theorem 1.12.7 (Gal [62]). *Suppose that* $1 \leq q < R$ *and that* $f : \mathbb{D}_R \to \mathbb{H}$ *is left W-analytic in* $\mathbb{D}_R$, *i.e.,* $f(z) = \sum_{p=0}^{\infty} a_p z^p$, *for all* $z \in \mathbb{D}_R$, *where* $a_p \in \mathbb{H}$ *for all* $p = 0, 1, 2, \ldots,$. *Also, denote* $S_k^{(q)} = [1]_q + \ldots + [k-1]_q$, $k \geq 2$:

(i) *If* $q = 1$, *then for any* $1 \leq r < R$, $\|z\| \leq r$ *and* $n \in \mathbb{N}$, *the following upper estimate*

$$\left\| B_{n,1}(f) - f - \sum_{k=2}^{\infty} a_k \cdot \frac{S_k^{(1)}}{n} [e_{k-1} - e_k] \right\| \leq \frac{1}{n^2} \cdot \sum_{k=2}^{\infty} \|a_k\| (k-1)^4 r^k,$$

holds, where $e_p(z) = z^p$ *and* $\sum_{k=2}^{\infty} \|a_k\| (k-1)^4 r^k < \infty$.

(ii) *If* $q > 1$, $1 < r < \frac{R}{q^2}$, $\|z\| \leq r$ *and* $n \in \mathbb{N}$, *then the following upper estimate*

$$\left\| B_{n,q}(f) - f - \sum_{k=2}^{\infty} a_k \cdot \frac{S_k^{(q)}}{[n]_q} [e_{k-1} - e_k] \right\| \leq \frac{C_{r,q}(f)}{[n]_q^2}$$

holds, where $C_{r,q}(f) = 4 \cdot \max \left\{ \frac{1}{(q-1)^2}, \frac{1}{(q-1)^3} \right\} \cdot \sum_{k=2}^{\infty} \|a_k\| (k-1)^2 (q^2 r)^k.$

(iii) If $q \geq 1$, then for any $1 < r < \frac{R}{q}$, we have

$$\lim_{n \to \infty} [n]_q (B_{n,q}(f)(z) - f(z)) = A_q(f)(z),$$

uniformly in $\overline{\mathbb{D}_r}$, where $A_q(f)(z) = \sum_{k=2}^{\infty} a_k \cdot S_k^{(q)} \cdot [z^{k-1} - z^k]$, $z \in \mathbb{H}$.

Proof. Let $1 < r < \frac{R}{q}$. Firstly, let us present the most important relationships which will be used in our proof. Simple calculation shows that

$$S_k^{(q)} = \frac{k(k-1)}{2}, \text{ for } q = 1 \text{ and } S_k^{(q)} = \frac{q^k - k(q-1) - 1}{(q-1)^2}, \text{ for } q > 1.$$

$$(1.12.1)$$

Also, reasoning exactly as in the proof of Theorem 2.3 in [60], we easily get

$$B_{n,q}(f)(z) = \sum_{k=0}^{\infty} a_k B_{n,q}(e_k)(z). \tag{1.12.2}$$

Here, note that by Theorem 2.1 in [60], we have

$$B_{n,q}(e_k)(z) = \sum_{p=0}^{n} \binom{n}{p}_q [\Delta_{1/[n]_q}^p e_k(0)]_q e_p(z) = \sum_{p=0}^{n} \binom{n}{p}_q [\Delta_{1/[n]_q}^p e_k(0)]_q e_p(z),$$

where

$$[\Delta_h^p f(0)]_q = \sum_{k=0}^{p} (-1)^q q^{k(k-1)/2} \binom{p}{k}_q f([p-k]_q h).$$

By the relationships (5), (6), and (7) in Ostrovska [118], p. 236 (see also relation (1), proof of Theorem 2.3 in [60]), it follows

$$B_{n,q}(e_k)(z) = \sum_{p=0}^{n} D_{n,p,k}^{(q)} e_p(z),$$

where

$$D_{n,p,k}^{(q)} = \binom{n}{p}_q \cdot q^{p(p-1)/2} \cdot [0, [1]_q/[n]_q, \ldots, [p]_q/[n]_q; e_k] \cdot \frac{[p]_q!}{[n]_q^p}$$

$$= \lambda_{n,p}^{(q)} \cdot [0, [1]_q/[n]_q, \ldots, [p]_q/[n]_q; e_k], \tag{1.12.3}$$

with $\lambda_{n,p}^{(q)} = \left(1 - \frac{[1]_q}{[n]_q}\right) \cdot \ldots \cdot \left(1 - \frac{[p-1]_q}{[n]_q}\right)$.

Note that by (1.12.3) (see also Lemma 3, p. 245 in [118]), we obtain

$$D_{n,p,k}^{(q)} \geq 0 \text{ for all } 0 \leq p \leq n, \ k \geq 0, \text{ and } \sum_{p=0}^{n} D_{n,p,k}^{(q)} = 1, \text{ for all } 0 \leq k \leq n$$

$$(1.12.4)$$

and

$$D_{n,k,k}^{(q)} = \Pi_{i=1}^{k-1}\left(1 - \frac{[i]_q}{[n]_q}\right), \quad D_{n,k-1,k}^{(q)} = \frac{S_k^{(q)}}{[n]_q} \cdot \Pi_{i=1}^{k-2}\left(1 - \frac{[i]_q}{[n]_q}\right), \quad k \leq n.$$

$$(1.12.5)$$

In what follows, first we prove that $A_q(f)(z)$ given by

$$A_q(f)(z) = \sum_{k=2}^{\infty} a_k \cdot S_k^{(q)} \cdot [z^{k-1} - z^k],$$

is left W-analytic in $\overline{\mathbb{D}_r}$, for $1 < r < \frac{R}{q}$.

Indeed, by the inequality

$$\|A_q(f)(z)\| \leq \sum_{k=0}^{\infty} \|a_k\| \cdot S_k^{(q)} \cdot [\|z^{k-1}\| + \|z^k\|],$$

since by (1.12.1) we get $S_k^{(q)} \leq \frac{q^k}{(q-1)^2}$ for $q > 1$, it immediately follows

$$\|A_q(f)(z)\| \leq \frac{2}{(q-1)^2} \cdot \sum_{k=0}^{\infty} \|a_k\|(qr)^k < \infty, \quad \text{if } q > 1,$$

and

$$\|A_q(f)(z)\| \leq \sum_{k=0}^{\infty} \|a_k\| k(k-1)(qr)^k < \infty, \quad \text{if } q = 1,$$

for all $z \in \overline{\mathbb{D}_r}$. These immediately show that for $q \geq 1$, the function $A_q(f)$ is well defined and left W-analytic in $\overline{\mathbb{D}_r}$.

Now, by (1.12.2) we obtain

$$\left\| B_{n,q}(f)(z) - f(z) - \sum_{k=0}^{\infty} a_k \cdot \frac{S_k^{(q)}}{[n]_q}[z^{k-1} - z^k] \right\|$$

$$\leq \sum_{k=0}^{\infty} \|a_k\| \cdot \|E_{k,n}^{(q)}(z)\|,$$

where

$$E_{k,n}^{(q)}(z) = B_{n,q}(e_k)(z) - z^k - \frac{S_k^{(q)}}{[n]_q}[z^{k-1} - z^k].$$

Because simple calculations and (1.12.1) imply that

$$E_{0,n}^{(q)}(z) = E_{1,n}^{(q)}(z) = E_{2,n}^{(q)}(z) = 0,$$

in fact we have to estimate the expression

$$\sum_{k=3}^{\infty} \|a_k\| \cdot \|E_{k,n}^{(q)}(z)\| = \sum_{k=3}^{n} \|a_k\| \cdot \|E_{k,n}^{(q)}(z)\| + \sum_{k=n+1}^{\infty} \|a_k\| \cdot \|E_{k,n}^{(q)}(z)\|.$$

To estimate $\|E_{k,n}^{(q)}(z)\|$, we distinguish two cases: (i) $3 \leq k \leq n$; (ii) $k \geq n+1$.

Case 1. We obtain

$$[n]_q \|E_{k,n}^{(q)}(z)\| = \|[n]_q (B_{n,q}(e_k)(z) - z^k) - S_k^{(q)} \cdot (z^{k-1} - z^k)\|$$

$$\leq r^k [n]_q \sum_{i=1}^{k-2} D_{n,i,k}^{(q)} + |[n]_q D_{n,k-1,k}^{(q)} - S_k^{(q)}| r^k + |[n]_q (1 - D_{n,k,k}^{(q)}) - S_k^{(q)}| r^k.$$

Taking now into account (1.12.4) and (1.12.5) and following exactly the reasonings in the proof of Lemma 3, p. 747 in [145], we arrive at

$$\|E_{k,n}^{(q)}(z)\| \leq \frac{4(k-1)^2 [k-1]_q^2}{[n]_q^2} \cdot r^k, \text{ for all } z \in \overline{\mathbb{D}_r}. \qquad (1.12.6)$$

By (1.12.6) it follows

$$\sum_{k=3}^{n} \|a_k\| \cdot \|E_{k,n}^{(q)}(z)\| \leq \frac{4}{[n]_q^2} \cdot \sum_{k=3}^{n} \|a_k\| (k-1)^2 [k-1]_q^2 r^k, \qquad (1.12.7)$$

for all $z \in \overline{\mathbb{D}_r}$ and $n \in \mathbb{N}$.

Case 2. We get

$$\sum_{k=n+1}^{\infty} \|a_k\| \cdot \|E_{k,n}^{(q)}(z)\| \leq \sum_{k=n+1}^{\infty} \|a_k\| \cdot \|B_{n,q}(e_k)(z)\| + \sum_{k=n+1}^{\infty} \|a_k\| \cdot \|z^k\|$$

$$+ \frac{1}{[n]_q} \sum_{k=n+1}^{\infty} \|a_k\| \cdot S_k^{(q)} \cdot \|z^{k-1}\| + \frac{1}{[n]_q} \sum_{k=n+1}^{\infty} \|a_k\| \cdot S_k^{(q)} \cdot \|z^k\|$$

$$=: L_{1,q}(z) + L_{2,q}(z) + L_{3,q}(z) + L_{4,q}(z). \qquad (1.12.8)$$

We have two subcases: (2_i) $q = 1$; (2_{ii}) $q > 1$.
Subcase (2_i). By (1.12.4), it immediately follows

$$L_{1,1}(z) \leq \sum_{k=n+1}^{\infty} \|a_k\| \cdot \sum_{p=0}^{n} D_{n,p,k}^{(1)} \|e_p(z)\| \leq \frac{1}{n^2} \sum_{k=n+1}^{\infty} \|a_k\| (k-1)^2 r^k,$$

and similarly

$$L_{2,1}(z) \leq \frac{1}{n^2} \sum_{k=n+1}^{\infty} \|a_k\|(k-1)^2 r^k.$$

for all $z \in \overline{\mathbb{D}_r}$.

Next, by similar reasonings as above and by (1.12.1) and (1.12.4), we obtain

$$L_{3,1}(z) \leq \frac{1}{n} \sum_{k=n+1}^{\infty} \|a_k\| \cdot \frac{k(k-1)}{2} r^k \leq \frac{1}{n^2} \sum_{k=n+1}^{\infty} \|a_k\| \cdot (k-1)^3 r^k,$$

and

$$L_{4,1}(z) \leq \frac{1}{n^2} \sum_{k=n+1}^{\infty} \|a_k\| \cdot (k-1)^3 r^k,$$

which by (1.12.8) implies that

$$\sum_{k=n+1}^{\infty} \|a_k\| \cdot \|E_{k,n}^{(1)}(z)\| \leq \frac{1}{n^2} \sum_{k=n+1}^{\infty} \|a_k\|(k-1)^3 r^k,$$

for all $z \in \overline{\mathbb{D}_r}$ and $n \in \mathbb{N}$.

But the sequence $\{a_n = \sum_{k=n+1}^{\infty} \|a_k\|(k-1)^3 r^k, n \in \mathbb{N}\}$ is convergent to zero (therefore bounded by a positive constant $M > 0$ independent of n), as the remainder of the convergent series $\sum_{k=0}^{\infty} \|a_k\|(k-1)^3 r^k$, which will imply

$$\sum_{k=n+1}^{\infty} \|a_k\| \cdot \|E_{k,n}^{(1)}(z)\| \leq \frac{M}{n^2} \leq \frac{\sum_{k=1}^{\infty} \|a_k\|(k-1)^3 r^k}{n^2},$$

for all $z \in \overline{\mathbb{D}_r}$ and $n \in \mathbb{N}$.

Now, taking $q = 1$ in (1.12.7) and taking into account that the series $\sum_{k=3}^{\infty} \|a_k\|(k-1)^4 r^k$ is convergent, we get

$$\sum_{k=3}^{n} \|a_k\| \cdot \|E_{k,n}^{(1)}(z)\| \leq \frac{4}{n^2} \cdot \sum_{k=3}^{n} \|a_k\|(k-1)^4 r^k$$

$$\leq \frac{4}{n^2} \cdot \sum_{k=3}^{\infty} \|a_k\|(k-1)^4 r^k,$$

which combined with the previous estimate immediately implies

$$\sum_{k=0}^{\infty} \|a_k\| \cdot \|E_{k,n}^{(1)}(z)\| \leq \frac{4 \sum_{k=2}^{\infty} \|a_k\|(k-1)^4 r^k}{n^2},$$

for all $z \in \overline{\mathbb{D}_r}$ and $n \in \mathbb{N}$.

This proves (i) in the statement of Theorem 1.12.7.

Subcase (2_{ii}). By (1.12.4), for all $z \in \overline{\mathbb{D}_r}$, it follows

$$L_{1,q}(z) \leq \sum_{k=n+1}^{\infty} \|a_k\| \cdot \sum_{p=0}^{n} D_{n,p,k}^{(q)} \|z^p\| \leq \sum_{k=n+1}^{\infty} \|a_k\| \cdot r^k$$

$$\leq \frac{1}{[n]_q^2} \sum_{k=n+1}^{\infty} \|a_k\| \cdot [k-1]_q^2 r^k \leq \frac{1}{(q-1)^2 [n]_q^2} \sum_{k=n+1}^{\infty} \|a_k\| \cdot (q^2 r)^k$$

and similarly

$$L_{2,q}(z) \leq \sum_{k=n+1}^{\infty} \|a_k\| \cdot r^k \leq \frac{1}{[n]_q^2} \sum_{k=n+1}^{\infty} \|a_k\| \cdot [k-1]_q^2 r^k$$

$$\leq \frac{1}{(q-1)^2 [n]_q^2} \sum_{k=n+1}^{\infty} \|a_k\| (q^2 r)^k.$$

Also, since by (1.12.1) we get $S_k^{(q)} \leq \frac{q^k}{(q-1)^2}$, for all $z \in \overline{\mathbb{D}_r}$, it follows

$$L_{3,q}(z) \leq \frac{1}{(q-1)^2 [n]_q} \sum_{k=n+1}^{\infty} \|a_k\| \cdot q^k r^k = \frac{1}{(q-1)^2 [n]_q} \sum_{k=n+1}^{\infty} \|a_k\| \cdot (qr)^k$$

$$\leq \frac{1}{(q-1)^3 [n]_q^2} \sum_{k=n+1}^{\infty} \|a_k\| (q^2 r)^k,$$

and similarly

$$L_{4,q}(z) \leq \frac{1}{[n]_q} \sum_{k=n+1}^{\infty} \|a_k\| \cdot q^k r^k \leq \frac{1}{(q-1)^3 [n]_q^2} \sum_{k=n+1}^{\infty} \|a_k\| \cdot (q^2 r)^k.$$

By (1.12.8), we immediately obtain

$$\sum_{k=n+1}^{\infty} |a_k| \cdot |E_{k,n}^{(q)}(z)| \leq \frac{C_{r,q}(f)}{[n]_q^2},$$

for all $z \in \overline{\mathbb{D}_r}$, where

$$C_{r,q}(f) = \max\left\{ \frac{1}{(q-1)^2}, \frac{1}{(q-1)^3} \right\} \cdot \sum_{k=n+1}^{\infty} \|a_k\| (q^2 r)^r.$$

Also, by (1.12.7), since $[k-1]_q^2 \leq [k]_q^2 \leq \frac{q^{2k}}{(q-1)^2}$, for $z \in \overline{\mathbb{D}_r}$ with $1 < r < \frac{R}{q^2}$, we easily obtain

$$\sum_{k=3}^{n} \|a_k\| \cdot \|E_{k,n}^{(q)}\| \leq \frac{4}{(q-1)^2[n]_q^2} \cdot \sum_{k=2}^{\infty} \|a_k\|(k-1)^2(q^2r)^k.$$

Collecting these results, we immediately obtain the upper estimate in (ii).

(iii) The case $q = 1$ follows directly by multiplying by n in the estimate in (i) and by passing to limit with $n \to \infty$. In the case of $q > 1$, if $1 < r < \frac{R}{q^2}$, then by multiplying in (ii) with $[n]_q$ and passing to limit with $n \to \infty$, we get the desired conclusion.

What remained to be proved is that the limit in (iii) still holds under the more general condition $1 < r < \frac{R}{q}$.

Since $\frac{R}{q^{1+t}} \nearrow \frac{R}{q}$ as $t \searrow 0$, evidently that given $1 < r < \frac{R}{q}$, there exists a $t \in (0,1)$, such that $q^{1+t}r < R$. Because f is left W-analytic in $\mathbb{D}_R$, this implies that $\sum_{k=2}^{\infty} \|a_k\| k^4 q^{(1+t)k} r^k = \sum_{k=2}^{\infty} \|a_k\| \cdot k^4 (q^{1+t}r)^k < \infty$, for all $z \in \mathbb{D}_r$. Also, the convergence of the previous series implies that for arbitrary $\varepsilon > 0$, there exists n_0, such that $\sum_{k=n_0+1}^{\infty} \|a_k\| \cdot k^2 q^k r^k < \varepsilon$. By using (1.12.6), for all $z \in \overline{\mathbb{D}_r}$ and $n > n_0$, we get

$$\|[n]_q(B_{n,q}(f)(z) - f(z)) - A_q(f)(z)\|$$

$$\leq \sum_{k=2}^{n_0} \|a_k\| \cdot \left\| [n]_q(B_{n,q}(e_k)(z) - e_k(z)) - S_k^{(q)}[z^{k-1} - z^k] \right\|$$

$$+ \sum_{k=n_0+1}^{\infty} \|a_k\| \cdot \left([n]_q \|B_{n,q}(e_k)(z) - z^k\| + S_k^{(q)} \|z^{k-1} - z^k\| \right)$$

$$\leq \sum_{k=2}^{n_0} \|a_k\| \cdot \frac{4(k-1)^2[k-1]_q^2}{[n]_q} \cdot r^k$$

$$+ \sum_{k=n_0+1}^{\infty} \|a_k\| \cdot \left([n]_q \|B_{n,q}(e_k)(z) - z^k\| + S_k^{(q)} \|z^{k-1} - z^k\| \right).$$

But by the proof of Theorem 2.3, Case 1 in [60], for $k \leq n$, we have

$$\|B_{n,q}(e_k)(z) - z^k\| \leq \frac{k[k-1]_q}{[n]_q} \cdot r^k,$$

while for $k > n$ and using (1.12.4), we get

$$\|B_{n,q}(e_k)(z) - z^k\| \leq \|B_{n,q}(e_k)(z)\| + \|z^k\| \leq \sum_{p=0}^{n} D_{n,p,k}^{(q)} \|z^p\| + \|z^k\|$$

$$\leq r^n + r^k \leq 2r^k \leq 2\frac{k[k-1]_q}{[n]_q} \cdot r^k,$$

for all $z \in \overline{\mathbb{D}_r}$.

Also, since $S_k^{(q)} \leq (k-1)[k-1]_q$, it is immediate that

$$S_k^{(q)} \cdot \|z^{k-1} - z^k\| \leq S_k^{(q)} \cdot [\|z^{k-1}\| + \|z^k\|] \leq 2(k-1)[k-1]_q r^k.$$

Therefore, we easily obtain

$$\sum_{k=n_0+1}^{\infty} \|a_k\| \cdot \left([n]_q \|B_{n,q}(e_k)(z) - z^k\| + S_k^{(q)} \|z^{k-1} - z^k\| \right)$$

$$\leq 2 \sum_{k=n_0+1}^{\infty} \|a_k\| \cdot (2k-1)[k-1]_q r^k,$$

valid for all $z \in \overline{\mathbb{D}_r}$.

In conclusion, for all $z \in \overline{\mathbb{D}_r}$ and $n > n_0$, we have

$$\|[n]_q(B_{n,q}(f)(z) - f(z)) - A_q(f)(z)\| \leq \sum_{k=2}^{n_0} \|a_k\| \cdot \frac{4(k-1)^2[k-1]_q^2}{[n]_q} \cdot r^k$$

$$+ 2 \sum_{k=n_0+1}^{\infty} \|a_k\| \cdot (2k-1)[k-1]_q r^k$$

$$\leq \frac{4}{[n]_q^t} \cdot \sum_{k=2}^{n_0} \|a_k\| \cdot k^2[k-1]_q^{1+t} \cdot r^k + 4 \sum_{k=n_0+1}^{\infty} \|a_k\| \cdot k^2 q^k r^k$$

$$\leq \frac{4}{[n]_q^t} \cdot \sum_{k=2}^{\infty} \|a_k\| \cdot k^4 q^{(1+t)k} \cdot r^k + 4\varepsilon.$$

Now, since $\frac{4}{[n]_q^t} \to 0$ as $n \to \infty$ and $\sum_{k=2}^{\infty} \|a_k\| \cdot k^4 q^{(1+t)k} \cdot r^k < \infty$, for the given $\varepsilon > 0$, there exists an index n_1, such that $\frac{4}{[n]_q^t} \cdot \sum_{k=2}^{\infty} \|a_k\| \cdot k^4 q^{(1+t)k} \cdot r^k < \varepsilon$ for all $n > n_1$.

As a final conclusion, for all $n > \max\{n_0, n_1\}$ and $z \in \overline{\mathbb{D}_r}$, we get

$$\|[n]_q(B_{n,q}(f)(z) - f(z)) - A_q(f)(z)\| \leq 5\varepsilon,$$

which shows that

$$\lim_{n \to \infty} [n]_q(B_{n,q}(f)(z) - f(z)) = A_q(f)(z), \quad \text{uniformly in } \overline{\mathbb{D}_r}.$$

The theorem is proved. $\qquad\qquad\qquad\qquad\qquad\qquad\qquad\qquad\qquad\qquad\qquad\square$

As a consequence of Theorem 1.12.7, we get the exact order of approximation, by the following.

Corollary 1.12.8 (Gal [62]). *Under the hypothesis of Theorem 1.12.7, suppose that $R > q \geq 1$. If $1 < r < \frac{R}{q}$ and f is not a left polynomial of degree ≤ 1 in $\mathbb{D}_R$ (namely, f is not of the form $f(z) = a_0 + a_1 z$), then*

$$\||B_{n,q}(f) - f\||_r \sim \frac{1}{[n]_q}, \quad n \in \mathbb{N},$$

holds, where $\||f\||_r = \sup\{\|f(z)\|; \|z\| \leq r\}$ and the constants in the equivalence depend on f, r, and q but are independent of n.

Proof. Suppose that f is such that the approximation order in approximation by $B_{n,q}(f)$ is less than $\frac{1}{[n]_q}$, that is, $\||B_{n,q}(f) - f\||_r \leq M \frac{s_n}{[n]_q}$, for all $n \in \mathbb{N}$, where $s_n \to 0$ as $n \to \infty$. This would imply that $\lim_{n \to \infty} [n]_q \||B_{n,q}(f) - f\||_r = 0$.

Then, by Theorem 1.12.7, (iii), it would immediately follow that $A_q(f) = 0$ for all $z \in \overline{\mathbb{D}_r}$, where $A_q(f)(z)$ is defined in the statement of Theorem 1.12.7, (iii). But $A_q(f)(z) = 0$ for all $z \in \overline{\mathbb{D}_r}$ by simple calculation implies

$$2a_2 S_2^{(q)} z + \sum_{k=2}^{\infty} [S_{k+1}^{(q)} a_{k+1} - S_k^{(q)} a_k] z^k = 0, \; z \in \overline{\mathbb{D}_r}.$$

By the uniqueness of power series (which follows from the Theorem 2.7 in [70]), since by (1.12.1) it is clear that $S_k^{(q)} > 0$ for all $k \geq 2$, we would get that $a_2 = 0$ and

$$S_{k+1}^{(q)} a_{k+1} - S_k^{(q)} a_k = 0, \; \text{for all } k = 2, 3, \ldots, .$$

For $k = 2$, we easily get $a_3 = 0$, and taking above step by step $k = 3, 4, \ldots$, we easily would obtain that $a_k = 0$ for all $k \geq 2$.

Therefore we would get $f(z) = a_0 + a_1 z$ for all $z \in \overline{\mathbb{D}_r}$; it would follow that f would be a left polynomial of degree ≤ 1 in $\overline{\mathbb{D}_r}$, a contradiction with the hypothesis. In conclusion, if f is not a left polynomial of degree ≤ 1, then by the Theorems 1.12.4 (for $q > 1$) and 1.12.6 (for $q = 1$), it immediately follows that the approximation order is exactly $\frac{1}{[n]_q}$, which proves the corollary. $\square$

Remarks. 1) By the obvious inequalities $\frac{q-1}{q^n} \leq \frac{1}{[n]_q} \leq \frac{q}{q^n}$, for all $n \in \mathbb{N}$ and $q > 1$, it follows that $B_{n,q}(f)$ approximates f in the disk $\overline{\mathbb{D}_r}$, with the exact order $\frac{1}{q^n}$.

2) In the complex variable case, Corollary 1.12.8 becomes Corollary 1 in [145].

3) If f is right W-analytic, then replacing $A_q(f)(z)$ by $A_q^*(f)(z) = \sum_{k=2}^{\infty} S_k^{(q)} [z^{k-1} - z^k] a_k$ and $B_{n,q}(f)(z)$ by $B_{n,q}^*(f)(z)$, by similar reasonings one can prove that the estimates and conclusions in the Theorem 1.12.7 and Corollary 1.12.8 still remain valid. In Corollary 1.12.8 the concept of left polynomials must be replaced by that of right polynomials.

1.13 Notes and Open Problems

Note 1.13.1. Theorems 1.3.1, 1.7.7, and 1.11.4 and Corollary 1.11.5 appear for the first time here. Also, note that the proofs of Theorem 1.7.1, (i); Theorem 1.7.2; and Theorem 1.7.6 are explained here in more detail than the original proofs in Gal [52].

Note 1.13.2. From a very long list, some references concerning the Bernstein-type operators of one real variable, different from those considered in this Chapter, for which would be possible to develop similar results are, for example, Altomare and Mangino [7]; Altomare and Raşa [8]; Bleimann, Butzer, and Hahn [21]; Cimoca and Lupaş [26]; Leviatan [95]; Lupaş [99, 100, 102]; Lupaş and Müller [103]; Meyer and König and Zeller [109]; Moldovan, G. [113]; Raşa [126]; Soardi [132]; and Stancu [134].

Open Problem 1.13.3. The complex Lorentz polynomials in Sect. 1.7 are defined by

$$L_n(f)(x) = \sum_{k=0}^{n} \binom{n}{k} \frac{f^{(k)}(0)}{n^k} z^k, n \in \mathbb{N},$$

a formula valid in any compact disk of center at origin. Then according o the ideas and methods used in Sect. 1.3, the Lorentz–Faber polynomials attached to a compact set $G \subset \mathbb{C}$ will be given by the formula

$$\mathcal{L}_n(f; \overline{G})(z) = \sum_{k=0}^{n} \binom{n}{k} \frac{F^{(k)}(0)}{n^k} F_k(z), \ z \in \overline{G}, n \in \mathbb{N},$$

where $F_k(z)$ are the Faber polynomials attached to G and F is defined as in Sect. 1.3, the lines before the statement of Theorem 1.3.1.

Since it is known (see Lorentz [96], p. 44, formula (4)) that if we denote the matrix $M = (a_{n,j})_{n,j \in \{0,1,\ldots,\}}$, with $a_{n,j} = \frac{j}{n} \left(1 - \frac{1}{n}\right) \ldots \left(1 - \frac{j-1}{n}\right)$ if $0 \leq j \leq n$, $a_{n,j} = 0$ if $j > n$, then we can write

$$L_n(f)(z) = \sum_{j=0}^{n} a_{n,j} s_j(z) \text{ where } s_j(z) = \sum_{p=0}^{j} \frac{f^{(p)}(0)}{p!} z^p,$$

it follows that

$$\mathcal{L}_n(f; \overline{G})(z) = \sum_{j=0}^{n} a_{n,j} S_j(z), \ z \in \overline{G}, n \in \mathbb{N},$$

where $S_j(z) = \sum_{p=0}^{j} \frac{F^{(p)}(0)}{p!} F_p(z)$ is the nth partial sum of the Faber expansion of F.

Note here (see Lorentz [96], p. 44) that the matrix M defines a regular method of summation (for the definition of regularity, e.g., Lorentz [96], p. 37), with positive coefficients and the sum $\sum_{j=0}^{n} a_{n,j} = 1$.

Now, having as a model the Schurer–Faber polynomials, it remains as an open question, the approximation properties of the Lorentz–Faber polynomials $\mathcal{L}_n(f; \overline{G})(z)$.

Open Problem 1.13.4. Analogous question with that in the Open Problem 1.13.3 can be posed for q-Lorentz–Faber polynomials, $q > 1$, which must produce the approximation order $1/q^n$.

Note and Open Problem 1.13.5. It is well known that the q analogue for integral-type operators, like those of Kantorovich type or of Durrmeyer type (see, e.g., Gal–Gupta–Mahmudov [67]), is defined by replacing in their formulas, among others, the classical integral with the Jackson's q-integral. Since the q-integral is defined for $0 < q \leq 1$ only (see Jackson [87], also Andrews, Askey, and Roy [11]), as a consequence it follows that the order of approximation for the complex operators cannot be better than $\frac{1}{n}$. In order to define these q-integral operators for the case of $q > 1$ (when the essentially better order of approximation, $\frac{1}{q^n}$, is expected), we would need to get for these q-integral operators expressions involving the classical integral only. In the q-Szász–Kantorovich case, the problem was solved in Gal, Mahmudov, and Kara [68], by defining $e_q(z) = \Pi_{j=0}^{\infty}(1 + (q-1)z/q^{j+1})$, $q > 1$,

$$
K_{n,q}(f, z) = \sum_{j=0}^{\infty} e_q(-[n]_q q^{-j} z) \cdot \frac{([n]_q z)^j}{[j]_q!} \cdot \frac{1}{q^{j(j+1)/2}} \int_0^1 f\left(\frac{q[j]_q + t}{[n+1]_q}\right) dt,
$$

and proving that if the nonconstant function $f : \mathbb{D}_R \bigcup [R, +\infty) \to \mathbb{C}$ is analytic in $\mathbb{D}_R$, continuous and bounded on $\overline{\mathbb{D}_R} \bigcup [R, +\infty)$, then for any $1 \leq r < \frac{R}{2q}$, we have $\|K_{n,q}(f) - f\|_r \sim \frac{1}{q^n}$.

Also, in the q-Bernstein–Kantorovich case, $q > 1$, the problem was solved in Mahmudov and Kara [106], by defining $p_{n,j}(q; z) = \binom{n}{j}_q z^j \Pi_{k=0}^{n-j-1}(1 - q^k z)$,

$$
K_{n,q}^*(f, z) = \sum_{j=0}^{n} p_{n,j}(q; z) \int_0^1 f\left(\frac{q[j]_q + t}{[n+1]_q}\right) dt,
$$

and proving that if the nonconstant function $f : \mathbb{D}_R \to \mathbb{C}$ is analytic in $\mathbb{D}_R$ and continuous in $\overline{\mathbb{D}_R}$, then we have $\|K_{n,q}^*(f) - f\|_r \sim \frac{1}{q^n}$, for any $1 \leq r < \frac{R}{q^2}$.

How to define and study complex q-Durrmeyer-kind operators for $q > 1$ remains an interesting open question.

Chapter 2

Overconvergence and Convergence in $\mathbb{C}$ of Some Integral Convolutions

This chapter deals with the overconvergence and convergence in $\mathbb{C}$ of some trigonometric convolution operators and with the approximation by some special type of convolutions called complex potentials, generated by the Beta and Gamma functions.

2.1 Complex Convolutions with Trigonometric-Type Kernels

In this section we study the overconvergence and convergence in $\mathbb{C}$ of the convolution operators based on trigonometric-type kernels.

2.1.1 Convolutions with Positive Trigonometric Kernels

This subsection deals with quantitative estimates in the overconvergence phenomenon for the classical convolution operators with positive trigonometric kernels. Also, in the particular cases of the Beatson kernel and their iterates, new shape-preserving properties are presented.

It is well known that if $f : \mathbb{R} \to \mathbb{R}$ is continuous and 2π-periodic and if $K_n(t) = \frac{1}{2} + \sum_{k=1}^{m_n} \rho_{k,n} \cos(kt)$ is a positive, even kernel (with $\rho_{k,n} \in \mathbb{R}$, for all k, n), then one can define the sequence of convolution trigonometric polynomials:

$$P_n(f)(x) = \frac{1}{\pi} \int_{-\pi}^{\pi} f(x+t)K_n(t)dt = \frac{1}{\pi} \int_{-\pi}^{\pi} f(u)K_n(u-x)du, \ n \in \mathbb{N}. \quad (2.1.1)$$

S.G. Gal, *Overconvergence in Complex Approximation*, 117
DOI 10.1007/978-1-4614-7098-4_2, © Springer Science+Business Media New York 2013

Classical choices for $K_n(t)$ are the de la Vallée–Poussin kernel, the Fejér kernel, the Jackson kernel, the generalized Jackson kernel, the Beatson kernel, and the Korovkin kernel, to mention only a few.

The quantitative convergence properties to $f(t)$ (of real variable $t \in \mathbb{R}$) of the above sequence $P_n(f)(t)$, $n = 1, 2, \ldots$, are well studied and can be found in any classical book in approximation theory (see, e.g., Stepanets [135] or DeVore–Lorentz [31]).

Supposing now that f is of complex variable $z \in \mathbb{C}$, the "complexification" of the above type of convolution polynomials can be done in two directions:

1) One replaces $x + t$ in the first expression in (2.1.1) of $P_n(f)$ by ze^{it} (here $i^2 = -1$), obtaining

$$P_n(f)(z) = \frac{1}{\pi} \int_{-\pi}^{\pi} f(ze^{it})K_n(t)dt,$$

which in the case when f is analytic in a disk D_r centered at origin, $P_n(f)(z)$ becomes a polynomial of degree $\leq m_n$ of complex variable $z \in D_r$. In this direction, the approximation (and geometric) properties of the sequence $(P_n(f)(z))_{n\in\mathbb{N}}$ were intensively studied in Chap. 3, Sect. 3.1, pp. 181–204 of the book of Gal [49], where even exact estimates were obtained.

2) One replaces x from $x+t$ simply by z in the first form in (2.1.1) of $P_n(f)x)$, obtaining

$$P_n(f)(z) = \frac{1}{\pi} \int_{-\pi}^{\pi} f(z+t)K_n(t)dt,$$

where we have to suppose that f is, for example, at least continuous in a strip of $\mathbb{C}$. In this case, evidently that $P_n(f)(z)$ loses its convolution character, but it would be of interest to study the overconvergence properties of $P_n(f)(z)$ in that strip containing the real axis.

The first goal of the present section is to study the above direction 2). It is easy to observe that if we replace x by z in the second form for $P_n(f)(x)$ in (2.1.1), then we don't obtain an operator with good approximation properties.

Clearly that the approximation properties of $P_n(f)(z)$ depend on the kernel.

For example, we can prove the following local/pointwise estimates.

Theorem 2.1.1. *Let $d > 0$ and suppose that $f : S_d \to \mathbb{C}$ is bounded and uniformly continuous in the strip $S_d = \{z = x + iy \in \mathbb{C}; x \in \mathbb{R}, |y| \leq d\}$. Also, denote $P_n(f)(z) = \frac{1}{\pi} \int_{-\pi}^{\pi} f(z+t)K_n(t)dt$:*

(i) If $K_n(t)$ is the de la Vallée–Poussin kernel, that is, $K_n(t) = \frac{1}{2} \cdot \frac{(n!)^2}{(2n)!} (2\cos[t/2])^{2n}$, then

$$|P_n(f)(z) - f(z)| \leq 3\omega_1(f; 1/\sqrt{n})_{[z-\pi,z+\pi]}, \text{ for all } z \in S_d, n \in \mathbb{N},$$

where $[z - \pi, z + \pi] = \{(z - \pi)(1 - \lambda) + \lambda(z + \pi); \lambda \in [0, 1]\}$ *and for* $0 \le \delta \le \pi$

$$\omega_1(f; \delta)_{[z-\pi, z+\pi]} =$$

$$sup\{|f(u + t) - f(u)|; u, u + t \in [z - \pi, z + \pi], t \in \mathbb{R}, |t| \le \delta\}.$$

(ii) If $K_n(t)$ *is the Fejér kernel, that is,* $K_n(t) = \frac{1}{2n} \cdot \left(\frac{sin[nt/2]}{sin[t/2]}\right)^2$, *then*

$$|P_n(f)(z) - f(z)| \le M\omega_1(f; 1/n)_{[z-\pi, z+\pi]}, \text{ for all } z \in S_d, n \in \mathbb{N},$$

where $M > 0$ *is a constant independent of* n *and* f.

(iii) If $K_n(t)$ *is the Jackson kernel, that is,* $K_n(t) = \frac{3}{2n(2n^2+1)} \cdot \left(\frac{sin[nt/2]}{sin[t/2]}\right)^4$, *then*

$$|P_n(f)(z) - f(z)| \le M\omega_2(f; 1/n)_{[z-\pi, z+\pi]}, \text{ for all } z \in S_d, n \in \mathbb{N},$$

where $M > 0$ *is a constant independent of* n *and* f *and for* $0 \le \delta \le \pi$

$$\omega_2(f; \delta)_{[z-\pi, z+\pi]} =$$

$$sup\{|f(u+t) - 2f(u) + f(u-t)|; u, u-t, u+t \in [z-\pi, z+\pi], t \in \mathbb{R}, |t| \le \delta\}.$$

Proof. (i) We immediately get

$$|P_n(f)(z) - f(z)| \le \frac{1}{\pi} \int_{-\pi}^{\pi} |f(z + t) - f(z)| K_n(t) dt$$

$$\le \frac{1}{\pi} \int_{-\pi}^{\pi} \omega_1(f; |t|)_{[z-\pi, z+\pi]} K_n(t) dt$$

$$\le \omega_1(f; 1/\sqrt{n})_{[z-\pi, z+\pi]} \int_{-\pi}^{\pi} (1 + |t|\sqrt{n}) K_n(t) dt \le 3\omega_1(f; 1/\sqrt{n})_{[z-\pi, z+\pi]}.$$

(For the last inequality see, e.g., Gal [55], p. 427).

(ii) As above we arrive at the estimate

$$|P_n(f)(z) - f(z)| \le \frac{1}{\pi} \int_{-\pi}^{\pi} |f(z + t) - f(z)| K_n(t) dt$$

$$\le \frac{1}{\pi} \int_{-\pi}^{\pi} \omega_1(f; |t|)_{[z-\pi, z+\pi]} K_n(t) dt$$

$$\le \omega_1(f; 1/n)_{[z-\pi, z+\pi]} \int_{-\pi}^{\pi} (1 + |t|n) K_n(t) dt \le M\omega_1(f; 1/\sqrt{n})_{[z-\pi, z+\pi]},$$

taking into account that for the Fejér kernel, we have (see, e.g., Gaier [37], Theorem 1) $\int_{-\pi}^{\pi} n|t|K_n(t)dt < M < \infty$, with $M > 0$ independent of n.

(iii) Firstly we easily obtain

$$P_n(f)(z) - f(z) = \frac{1}{\pi} \int_0^{\pi} [f(z+t) - 2f(z) + f(z-t)]K_n(t)dt,$$

which immediately implies

$$|P_n(f)(z) - f(z)| = \frac{1}{\pi} \int_{-\pi}^{\pi} \omega_2(f;t)_{[z-\pi,z+\pi]}K_n(t)dt$$

$$\leq \frac{1}{\pi}\omega_2(f;1/n)_{[z-\pi,z+\pi]} \int_0^{\pi} (1+nt)^2 K_n(t)dt \leq M\omega_2(f;1/n)_{[z-\pi,z+\pi]},$$

because by, e.g., Lorentz [97], p. 56, we have $\int_0^{\pi}(1+nt)^2 K_n(t)dt \leq C$, with $C > 0$ independent of n.   □.

The second goal of the present section is to discuss the shape-preserving properties of the complex convolution operators as defined by the above-mentioned direction 1), in the particular cases of the Beatson kernel and their iterates. At the beginning, we will recall some approximation and shape-preserving properties of these complex convolutions based on the Beatson kernel and their iterates. For that purpose we need some preliminaries, as follows.

Let us consider the open disk $\mathbb{D}_R = \{z \in \mathbb{C}; |z| < R\}$ and

$$A(\mathbb{D}_1) = \{f : \overline{\mathbb{D}_1} \to \mathbb{C}; f \text{ is analytic on } \mathbb{D}_1, \text{ and } f(0) = 0, f'(0) = 1\}.$$

It is well known that $f \in A(\mathbb{D}_1)$ is called starlike if $f(\mathbb{D}_1)$ is a starlike plane domain with respect to 0 and it is called convex if $f(\mathbb{D}_1)$ is a convex plane domain.

Define now for $n, r \in \mathbb{N}$ the Beatson kernels by (see Beatson [13])

$$B_{n,r}(t) = \frac{n}{2\pi c_{n,r}} \int_{t-\pi/n}^{t+\pi/n} K_{n,r}(s)ds,$$

where $K_{n,r}(t)$ are the Jackson kernels given by

$$K_{n,r}(s) = \left(\frac{sin\frac{ns}{2}}{sin\frac{s}{2}}\right)^{2r},$$

with $c_{n,r}$ chosen such that $\frac{1}{\pi}\int_{-\pi}^{\pi} K_{n,r}(s)ds = c_{n,r}$, and the iterates of the Beatson kernels by (see Gal [40]) $B_{n,r,1}(t) := B_{n,r}(t)$,

$$B_{n,r,2}(t) = \frac{n}{2\pi} \int_{t-\pi/n}^{t+\pi/n} B_{n,r,1}(s)ds \quad ,\dots,$$

$$B_{n,r,p}(t) = \frac{n}{2\pi} \int_{t-\pi/n}^{t+\pi/n} B_{n,r,p-1}(s)ds,$$

$p = 2, 3 \dots,, n, r \in \mathbb{N}$.

Through these trigonometric kernels, one can define the complex convolutions

$$L_{n,r}(f)(z) = \frac{1}{\pi} \int_{-\pi}^{\pi} f(ze^{iu})B_{n,r}(u)du$$

and

$$L_{n,r,p}(f)(z) = \frac{1}{\pi} \int_{-\pi}^{\pi} f(ze^{iu})B_{n,r,p}(u)du.$$

Remark. The approximation properties by these complex convolutions were obtained in terms of the second-order modulus of smoothness in Gal [39], at p. 243 and p. 246.

The following subclasses of functions are important in geometric function theory:

$$S_1 = \{f : \mathbb{D}_1 \to \mathbb{C}; f(z) = z + a_2 z^2 + \dots, \text{ analytic in } \mathbb{D}_1 \text{ and } \sum_{k=2}^{\infty} k|a_k| \le 1\},$$

$$S_2 =$$
$$= \{f : \overline{\mathbb{D}_1} \to \mathbb{C}; f \text{ is analytic on } \mathbb{D}_1, f(0) = f'(0)-1 = 0, |f''(z)| \le 1, z \in \mathbb{D}_1\},$$

$$\mathcal{R} =$$
$$= \{f : \overline{\mathbb{D}_1} \to \mathbb{C}; f \text{ is analytic on } \mathbb{D}, f(0) = f'(0)-1 = 0, Re f'(z) > 0, z \in \mathbb{D}_1\},$$

and

$$S_M =$$
$$= \{f : \overline{\mathbb{D}_1} \to \mathbb{C}; f \text{ is analytic on } \mathbb{D}, f(0) = f'(0)-1 = 0, |f'(z)| < M, z \in \mathbb{D}_1\}.$$

According to, e.g., Mocanu, Bulboacă, and Sălăgean [110], p. 97, Exercise 4.9.1, if $f \in S_1$, then $|\frac{zf'(z)}{f(z)} - 1| < 1, z \in \mathbb{D}_1$ and therefore f is starlike (and univalent) on $\mathbb{D}_1$.

By Obradović [116] it follows that $f \in S_2$ implies that f is starlike univalent on $\mathbb{D}_1$.

Also, it is known that $\mathcal{R}$ is called the class of functions with bounded turn (because $f \in \mathcal{R}$ is equivalent to $|arg f'(z)| < \frac{\pi}{2}, z \in \mathbb{D}_1$) and that $f \in \mathcal{R}$ implies the univalence of f on $\mathbb{D}_1$.

Finally, according to, e.g., Mocanu, Bulboacă, and Sălăgean [110], p. 111, Exercise 5.4.1, $f \in S_M$ implies that f is univalent on $\mathbb{D}_{\frac{1}{M}} = \{z \in \mathbb{C}; |z| < \frac{1}{M}\}$.

We can present the following shape-preserving properties, recalled in Gal and Greiner [65] too:

Theorem 2.1.2. *Let $n, r, p \in \mathbb{N}$.*

(i) *(see also Gal [40], Theorem 1, (ii)) If f is convex on $\mathbb{D}_1$, then $L_{n,r,p}(f)$ is close to convex;*

(ii) *(see also Gal [42], Theorem 3.3, (i)) $L_{n,r,p}(S_1) \subset S_1$;*

(iii) *(see also Gal [42], Theorem 3.4, (i)) $L_{n,r,p}(S_2) \subset S_2$ and $L_{n,r,p}(\mathcal{R}) \subset \mathcal{R}$;*

(iii) *(see also [42], Theorem 3.5) $L_{n,r,p}(S_M) \subset S_M$.*

Unfortunately, the convolution $L_{n,r}(f)(z)$ does not preserve the convexity of f. More exactly we have:

Theorem 2.1.3 (Gal–Greiner [65]). *Let $f \in A(\mathbb{D})$. The convolution polynomial defined by*

$$L_{n,r}(f)(z) = \frac{1}{\pi} \int_{-\pi}^{\pi} f(ze^{iu}) B_{n,r}(u) du,$$

$z = re^{ix} \in \mathbb{D}$, *does not preserve the convexity of f for any $n, r \in \mathbb{N}$.*

Proof. Take, for example, the Köebe function $k(z) = \frac{z}{(1-z)^2}$. Then by straightforward calculation, we get that up to a constant we have

$$L_{n+1,1}(k)(z) = \sum_{k=0}^{n} \frac{\sin(k\pi/(n+1))}{\sin(\pi/(n+1))} \cdot \frac{n+1-k}{n+1} \cdot z^k.$$

These polynomials are known to be univalent but not convex in any direction; see Suffridge [137].

Also, plots of $L_{n,r}(k)(z)$, for $r \geq 2$ and $n \in \mathbb{N}$ lead (at least numerically) to polynomials which are not convex in any direction. $\qquad\square$

In what follows, connected to a famous problem of Schoenberg, we will give an explanation of the negative result contained by Theorem 2.1.3.

Thus, if $Q(t)$ is a 2π-periodic kernel, $f \in A(\mathbb{D}_1)$ and one defines the convolution

$$L(f)(z) = \frac{1}{2\pi} \int_{-\pi}^{\pi} f(ze^{it}) Q(t) dt,$$

then in Ruscheweyh and Salinas [127], the complete solution to the problem in Schoenberg [129] is found, by proving the following result.

Theorem 2.1.4 (Ruscheweyh–Salinas [127]). *The convolution defined through the 2π-periodic kernel $Q(t)$ as above (with $Q(t)$ appropriate smooth) is convexity preserving if and only if the following conditions are satisfied:*

(i) $Q(t)$ is periodically monotone, that is, monotonically increasing on the first subinterval of $[-\pi, \pi]$ and decreasing on the second subinterval of $[-\pi, \pi]$.

(ii) $\log |Q'(t)|$ is concave in every subinterval of $\mathbb{R}$ in which $Q(t)$ does not take its minimum or maximum value.

Remark. It is easy to see that for an appropriate smooth kernel $Q(t)$, the fact that $\log |Q'(t)|$ is concave, one reduces to the inequality

$$[Q''(t)]^2 - Q'(t) \cdot Q'''(t) \geq 0, \text{ for all } t \in \mathbb{R}. \tag{A}$$

The main result connected to the Beatson kernels is the following.

Theorem 2.1.5 (Gal–Greiner [65]). *For any $n, r \in \mathbb{N}$ the nth Beatson kernel of order r, $B_{n,r}(t)$, satisfies*

$$-B'_{n,r}(t) = B'_{n,r}(-t) \geq 0, \text{ for all } t \in [0, \pi], \tag{2.1.2}$$

$$B''_{n,r}(t)^2 - B'_{n,r}(t)B'''_{n,r}(t) \geq 0, \text{ for all } t \in \mathbb{R}. \tag{2.1.3}$$

Proof. The case $n = 1$ is immediate. Fix $n, r \in \mathbb{N}$, $n \geq 2$, and let

$$b(t) := \frac{2\pi}{nc_{n,r}} B'_{n,r}(t) = \frac{1}{c_{n,r}} \left(K_{n,r}\left(t + \frac{\pi}{n}\right) - K_{n,r}\left(t - \frac{\pi}{n}\right) \right)$$

$$= \left(\frac{\cos \frac{nt}{2}}{\sin \frac{nt+\pi}{2n}} \right)^{2r} - \left(\frac{\cos \frac{nt}{2}}{\sin \frac{nt-\pi}{2n}} \right)^{2r}$$

$$= \left(\frac{2 \cos \frac{nt}{2}}{\cos \frac{\pi}{n} - \cos t} \right)^{2r} \left(\left(\sin \frac{nt - \pi}{2n} \right)^{2r} - \left(\sin \frac{nt + \pi}{2n} \right)^{2r} \right)$$

$$= f(t)^{2r} g(t),$$

where

$$f(t) := \frac{2 \cos \frac{nt}{2}}{\cos \frac{\pi}{n} - \cos t},$$

$$g(t) := \left(\sin \frac{nt - \pi}{2n} \right)^{2r} - \left(\sin \frac{nt + \pi}{2n} \right)^{2r}.$$

(Observe that $f(t)^2$ is up to normalization the nth Fejér–Korovkin kernel.)

Inequality (2.1.2) now immediately follows from $g(t) \leq 0$ for $t \in [0, \pi]$. For the proof of (2.1.3) we are left to show, because of symmetry and periodicity, that

$$0 \leq b'(t)^2 - b(t)b''(t)$$
$$= f(t)^{4r-2} \left[2r \left(f'(t)^2 - f(t)f''(t) \right) g(t)^2 + f(t)^2 \left(g'(t)^2 - g(t)g''(t) \right) \right].$$

for all $t \in [0, \pi]$. To this end we shall establish the two inequalities

$$0 \leq f'(t)^2 - f(t)f''(t), \tag{2.1.4}$$

$$0 \leq g'(t)^2 - g(t)g''(t). \tag{2.1.5}$$

Inequality (2.1.5) is implied by the representation

$$g'(t)^2 - g(t)g''(t) = \frac{r}{2}\left[\left(\sin\frac{nt+\pi}{2n}\right)^{2r-1} + \left(\sin\frac{nt-\pi}{2n}\right)^{2r-1}\right]^2$$

$$+r\left(2r\cos\frac{\pi}{n} + 2r - 1 - \cos t\right)\left(\sin\frac{\pi}{2n}\right)^2\left(\sin\frac{nt+\pi}{2n}\right)^{2r-2}\left(\sin\frac{nt-\pi}{2n}\right)^{2r-2}$$

$$\geq 0.$$

To prove inequality (2.1.4) we first observe

$$\left(\cos t - \cos\frac{\pi}{n}\right)^4 [f'(t)^2 - f(t)f''(t)]$$

$$= n^2\left(\cos t - \cos\frac{\pi}{n}\right)^2 - 2\left(1 - \cos t\cos\frac{\pi}{n}\right)(1 + \cos nt)$$

$$= n^2\left(x - \cos\frac{\pi}{n}\right)^2 - 2\left(1 - x\cos\frac{\pi}{n}\right)(1 + T_n(x)),$$

where $x = \cos t \in [-1, 1]$ and T_n is the nth Chebyshev polynomial of the first kind. Hence we are left to show that

$$p_n(x) := 2\left(1 - x\cos\frac{\pi}{n}\right)(1 + T_n(x)) \leq n^2\left(\cos\frac{\pi}{n} - x\right)^2, \tag{2.1.6}$$

for $x \in [-1, 1]$. If $n = 2, 3$ is immediate, for $n \geq 4$, we consider two cases to prove (2.1.6). First, let $x \in [-1, \cos(2\pi/n)]$. Since $1 + T_n(x) = 1 + T_n(\cos t) = 1 + \cos nt \leq 2$ and

$$\frac{n^2}{2}\frac{(x - \cos\frac{\pi}{n})^2}{1 - x\cos\frac{\pi}{n}} = \frac{(1 + 2\cos\frac{\pi}{n})^2}{1 + 2\cos\frac{\pi}{n} + 2(\cos\frac{\pi}{n})^2}\left(n\sin\frac{\pi}{2n}\right)^2$$

$$\geq \frac{(1 + 2\cos\frac{\pi}{4})^2}{1 + 2\cos\frac{\pi}{4} + 2(\cos\frac{\pi}{4})^2}\left(n\frac{2\sqrt{2}}{\pi}\frac{\pi}{2n}\right)^2$$

$$= 2 + \sqrt{2} > 2$$

we obtain (2.1.6) for those x.

Now, let $x \in [\cos(2\pi/n), 1]$. Since $p_n(\cos(\pi/n)) = p'_n(\cos(\pi/n)) = 0$, inequality (2.1.6) follows from

$$p''_n(x) \leq 2n^2, \text{ for } x \in [\cos(2\pi/n), 1], \tag{2.1.7}$$

by integrating twice. Hence, we are left to establish (2.1.7).

Elementary properties of the Chebyshev polynomials T_n as their differential equation

$$(1 - x^2)T_n''(x) - xT_n'(x) + n^2 T_n(x) = 0$$

and

$$T_n\left(\cos\frac{k\pi}{n}\right) = (-1)^k \qquad \text{for } k = 0, 1, \ldots, n,$$

$$T_n'\left(\cos\frac{k\pi}{n}\right) = 0 \qquad \text{for } k = 1, 2 \ldots, n - 1,$$

imply

$$p_n''\left(\cos\frac{k\pi}{n}\right) = (-1)^{k+1} 2n^2 \frac{1 - \cos\frac{k\pi}{n}}{1 - (\cos\frac{k\pi}{n})^2} \qquad \text{for } k = 1, 2 \ldots, n - 1$$

and show together with $\lim_{x \to \infty} p_n''(x) = -\infty$ that this polynomial of exact degree $n - 1$ has one zero in the interval $(\cos(\pi/n), +\infty)$ and one zero in each of the $n - 2$ intervals $(\cos(k\pi/n), \cos((k-1)\pi/n))$, $k = 2, 3, \ldots, n - 1$. Hence, p_n'' has exactly $n - 2$ extremas $x_{n-2} < x_{n-1} < \ldots < x_1$, each one of them between two consecutive zeros. In this enumeration, the point x_l is a maximum if l is odd and a minimum if l is even. Particularly we have $x_3 < \cos(2\pi/n)$. Furthermore,

$$p_n^{(3)}\left(\cos\frac{\pi}{n}\right) = 0,$$

$$p_n^{(4)}\left(\cos\frac{\pi}{n}\right) = -\frac{n^2}{12}\frac{1}{(\sin\frac{\pi}{n})^2}\left(n^2 - 1 - \frac{3}{(\sin\frac{\pi}{n})^2}\right) < 0,$$

which imply $x_1 = \cos(\pi/n)$. Therefore, p_n'' attains its maximum in the interval $[\cos(2\pi/n), 1]$ at the point $\cos(\pi/n)$. This proves (2.1.7). $\qquad \square$

Remarks. 1) Inequality (2.1.6) also follows by writing $1 + T_n(x)$ as a product (knowing its zeros), dividing the left-hand side of (2.1.6) by the right-hand side and an examination of its logarithmic derivative in the interval $(\cos(3\pi/n), 1)$. This may shorten the proof a little bit, depending on how explicit one would like to be.

2) In conclusion, Theorem 2.1.5 combined with Theorem 2.1.3 show that despite of the fact that the kernels $B_{n,r}(t)$ satisfy condition (i) in Theorem 2.1.4 and the above inequality (A), however do not fully satisfy condition (ii) in Theorem 2.1.4, because any $B_{n,r}'(t)$ has roots at which $B_{n,r}(t)$ does not neither take its minimum nor its maximum value.

Notice that it is for the first time when one really has to read the characterization in Theorem 2.2.4 so carefully that the small difference between condition (ii) in Theorem 2.1.4 and the inequality (A) in the Remark after the statement of Theorem 2.1.4 comes up.

Finally, several interesting open problems could be raised, as follows:

Problem 1. Do the inequality (2.1.3) in Theorem 2.1.5 satisfy the iterative Beatson kernels $B_{n,r,p}(t)$ too?

Problem 2. If the answer to the Problem 1 is positive, then for $p \geq 2$, do the iterative Beatson kernels $B_{n,r,p}(t)$ satisfy for the conditions in Theorem 2.1.4?

Problem 3. Do the convolution polynomials $L_{n,r}(f)(z)$ or $L_{n,r,p}(f)(z)$ preserve the subordination and the distortion of $f(z)$?

Problem 4. What other geometric properties could the convolutions have based on the Beatson kernels?

2.1.2 Convolutions with Nonpositive Cosine Kernels

In this subsection we derive approximation properties of the complex convolution operators based on nonpositive kernels of the form

$$K_{p,t}(u) = \int_0^\infty e^{-s^p} \cos\left\{ \frac{us}{t^{1/(p)}} \right\} ds, \, p \in \mathbb{N}, t \in \mathbb{R}_+.$$

More exactly, we deal with the following two types of complex convolutions, which were defined and studied in Gal, Gal, and Goldstein [64]:

$$S_q(t)f(z) =$$

$$= \frac{1}{\pi t^{1/(2q)}} \int_{-\infty}^{+\infty} \left[\int_0^\infty e^{-s^{2q}} \cos\left\{ \frac{us}{t^{1/(2q)}} \right\} ds \right] f(ze^{-iu}) du, \, q \geq 2,$$

and

$$T_q(t)f(z) =$$

$$= \frac{1}{\pi t^{1/(2q+1)}} \int_{-\infty}^{+\infty} \left[\int_0^\infty e^{-s^{2q+1}} \cos\left\{ \frac{us}{t^{1/(2q+1)}} \right\} ds \right] f(ze^{-iu}) du, \, q \geq 1,$$

where $z \in \overline{\mathbb{D}_1}$, $t \geq 0$ and f is considered analytic in $\mathbb{D}_1$ and continuous in $\overline{\mathbb{D}_1}$. In this sense, we present the following.

Theorem 2.1.6 (Gal–Gal–Goldstein [64]). *Let f be analytic in $\mathbb{D}_1$ and continuous in $\overline{\mathbb{D}_1}$, and $t \in \mathbb{R}$, $t \geq 0$.*

(i) For $q \in \mathbb{N}$, $q \geq 2$, the following estimate holds:

$$|S_q(t)f(z) - f(z)| \leq C_{2q}\omega_1(f; t^{1/(2q)})_{\overline{\mathbb{D}_1}}, \text{ for all } z \in \overline{\mathbb{D}_1}, , \, t > 0,$$

where $C_{2q} > 0$ is a constant independent of t and f and $\omega_1(f; \delta)_{\overline{\mathbb{D}_1}}$ denotes the modulus of continuity, defined by

$$\omega_1(f; \delta)_{\overline{\mathbb{D}_1}} = \sup\{|f(u) - f(v)| : |u - v| \le \delta, u, v \in \overline{\mathbb{D}_1}\}.$$

(ii) For $q \in \mathbb{N}$, $q \ge 1$, we have

$$|T_q(t)f(z) - f(z)| \le C_{2q+1}\omega_1(f; t^{1/(2q+1)})_{\overline{\mathbb{D}_1}}, \text{ for all } z \in \overline{\mathbb{D}_1}, t > 0,$$

where $C_{2q+1} > 0$ is a constant independent of t and f.

Proof. (i) Reasoning exactly as in the proof of Theorem 2.1, (v) in Gal, Gal, and Goldstein [64] (whose details are too long to be reproduced here) and taking into account the maximum modulus theorem, we obtain

$$|S_q(t)f(z) - f(z)| \le \frac{1}{\pi} \int_{-\infty}^{+\infty} \left| \int_0^\infty e^{-ts^{2q}} \cos(\alpha s)ds \right| |f(ze^{-i\alpha}) - f(z)|d\alpha$$

$$\le \frac{1}{\pi} \int_{-\infty}^{+\infty} \left| \int_0^\infty e^{-ts^{2q}} \cos(\alpha s)ds \right| \omega_1(f; |1 - e^{-i\alpha}|)_{\overline{\mathbb{D}_1}} d\alpha$$

$$= \frac{1}{\pi} \int_{-\infty}^{+\infty} \left| \int_0^\infty e^{-ls^{2q}} \cos(\alpha s)ds \right| \omega_1\left(f; 2\left|\sin\frac{\alpha}{2}\right|\right)_{\overline{\mathbb{D}_1}} d\alpha$$

$$\le \frac{1}{\pi} \int_{-\infty}^{+\infty} \left| \int_0^\infty e^{-ts^{2q}} \cos(\alpha s)ds \right| \omega_1(f; |\alpha|)_{\overline{\mathbb{D}_1}} d\alpha$$

$$\le C_{2q}\omega_1(f; t^{1/(2q)})_{\overline{\mathbb{D}_1}}.$$

(ii) The proof is similar to that from the above point (i). $\square$

Remark. Denoting

$$A(\mathbb{D}_1) = \{f : \mathbb{D}_1 \to \mathbb{C}; f \text{ is analytic in } \mathbb{D}_1 \text{ and continuous on } \overline{\mathbb{D}_1}\},$$

in Gal, Gal, and Goldstein [64], it is proved that $(S_q(t), t \ge 0)$ and $(T_q(t), t \ge 0)$ are (C_0)-semigroups of linear operators on $A(\mathbb{D}_1)$ and $u_q(t, \cdot) = S_q(t)f(\cdot)$ and $v_q(t, \cdot) = T_q(t)f(\cdot)$ are the unique solutions of the following two Cauchy problems for higher-order evolution equations in $\mathbb{C}$:

$$\frac{\partial u_q}{\partial t}(t, z) = (-1)^{q+1}\frac{\partial^{2q} u_q}{\partial \varphi^{2q}}(t, z), (t, z) \in (0, +\infty) \times D, z = re^{i\varphi}, z \neq 0,$$

$$u(0, z) = f(z), z \in \overline{D}, f \in A(\mathbb{D}_1)$$

and

$$\frac{\partial^2 v_q}{\partial t^2}(t,z) + \frac{\partial^{2(2q+1)} v_q}{\partial \varphi^{2(2q+1)}}(t,z) = 0, \ (t,z) \in (0,+\infty) \times D, \ z = re^{i\varphi}, \ z \neq 0,$$

$$v_q(0,z) = f(z), \ z \in \overline{D}, \ f \in A(\mathbb{D}_1),$$

respectively.

2.2 Approximation by Complex Potentials of Euler Type

In the real case, the approximation properties of the potentials such as those of Riesz, Bessel, generalized Riesz, generalized Bessel, and Flett have been studied by many authors; see, e.g., Kurokawa [93]; Gadjiev, Aral, and Aliev [36]; Uyhan, Gadjiev, and Aliev [140]; Sezer [131]; Aliev, Gadjiev, and Aral [6]; and their references.

In this section, we obtain some results concerning the approximation by several types of complex potentials generated by the Γ and *Beta* Euler's functions.

Let us recall that in the real case, the classical Bessel-type potential is defined for any $f \in L^p(\mathbb{R}^2)$, $1 \leq p < \infty$, by

$$B^\alpha(f)(x,t) = \frac{1}{\Gamma(\alpha/2)} \int_0^\infty \left[\int_{-\infty}^\infty \tau^{(\alpha/2)-1} e^{-\tau} W(y,\tau) f(x-y, t-\tau) dy \right] d\tau,$$

where $\alpha > 0$, $\Gamma(\alpha)$ is the Gamma function and $W(y,\tau) = \frac{1}{\sqrt{4\pi\tau}} e^{-y^2/(4\tau)}$ is the Gauss–Weierstrass kernel.

It is known that formally, we can write

$$B^\alpha(f)(x,t) = \left(I - \frac{\partial^2}{\partial x^2} + \frac{\partial}{\partial t} \right)^{-\alpha/2} f(x,t),$$

and the following convergence properties hold (see Uyhan–Gadjiev–Aliev [140]):

(i) *If $f \in L^p(\mathbb{R}^2)$, $1 \leq p < \infty$, is continuous at $(x,t) \in \mathbb{R}^2$ then $\lim_{\alpha \to 0+} B^\alpha(f)(x,t) = f(x,t)$.*

(ii) *If $f \in L^p(\mathbb{R}^2) \cap C_0(\mathbb{R}^2)$, where $C_0(\mathbb{R}^2)$ denotes the space of all continuous functions on $\mathbb{R}^2$ vanishing at infinity, then $\lim_{\alpha \to 0+} B^\alpha(f) = f$ uniformly on $\mathbb{R}^2$.*

(iii) *If $f \in L^p(\mathbb{R}^2) \cap C(\mathbb{R}^2)$, where $C(\mathbb{R}^2)$ denotes the space of all continuous functions on $\mathbb{R}^2$, then $\lim_{\alpha \to 0+} B^\alpha(f) = f$ uniformly on every compact $K \subset \mathbb{R}^2$.*

(iv) *In addition, for f in some suitable Lipschitz-type classes, quantitative upper estimates of order $O(\alpha)$ are obtained.*

Also, let us recall that the classical Flett potential is defined for any $f \in L^p(\mathbb{R})$ by (see Flett [35])

$$F^\alpha(f)(x) = \frac{1}{\Gamma(\alpha)} \int_0^\infty t^{\alpha-1} e^{-t} Q_t(f)(x) dt,$$

where $Q_t(f)(x) = \frac{t}{\pi} \int_{-\infty}^\infty \frac{f(x-u)}{u^2+t^2} du$ is the classical Poisson–Cauchy singular integral.

It is known that the following convergence properties hold (see Sezer [131]):

(i) If $f \in L^p(\mathbb{R}) \cap C_0(\mathbb{R})$, then $\lim_{\alpha \to 0+} F^\alpha(f) = f$ uniformly on $\mathbb{R}$.;

(ii) For f in some suitable Lipschitz-type classes, quantitative upper estimates of order $O(\alpha)$ are obtained.

Remark. The form of the Flett potential suggests us to study the approximation properties as $\alpha \to 0^+$ of new potentials, as follows:

$$F_U^\alpha(f)(x) = \frac{1}{\Gamma(\alpha)} \int_0^\infty t^{\alpha-1} e^{-t} U_t(f)(x) dt,$$

where $U_t(f)(x)$ can be any from $P_t(f)(x) = \frac{1}{2t} \int_{-\infty}^{+\infty} f(x-u) e^{-|u|/t} du$ (the Picard singular integral), $R_t(f)(x) = \frac{2t^3}{\pi} \int_{-\infty}^{+\infty} \frac{f(x-u)}{(u^2+t^2)^2} du$ (a Poisson–Cauchy-type singular integral), and $W_t^*(f)(x) = \frac{1}{\sqrt{\pi t}} \int_{-\infty}^{+\infty} f(x-u) e^{-u^2/t} du$ (the Gauss–Weierstrass singular integral). Also, in the form of the Flett potential, we could replace the Gamma function with other special function, for example, with the Beta function, so that we could study the approximation properties as $\alpha \to 0^+$, of new potentials of the form

$$G_U^{\alpha,\beta}(f)(x) = \frac{1}{Beta(\alpha,\beta)} \int_0^1 t^{\alpha-1}(1-t)^{\beta-1} U_t(f)(x) dt,$$

where $\alpha, \beta > 0$, $\alpha + \beta \geq 1$ and $U_t(f)(x)$ are any of the above-mentioned singular integrals.

In what follows, first we study the approximation properties of the complex versions of the potentials $F_U^\alpha(f)(x)$ (that includes the Flett potential) and $G_U^\alpha(f)(x)$. The complexification is made in two directions:

1) The complex forms are obtained from their real versions by replacing the translation $x - y$ by the rotation ze^{-iy}, where $z = re^{ix} \in \mathbb{C}$, that is, in the convolution form

$$F_U^\alpha(f)(z) = \frac{1}{\Gamma(\alpha)} \int_0^\infty t^{\alpha-1} e^{-t} U_t(f)(z) dt,$$

$$G_U^{\alpha,\beta}(f)(z) = \frac{1}{Beta(\alpha,\beta)} \int_0^1 t^{\alpha-1}(1-t)^{\beta-1} U_t(f)(z) dt,$$

where $U_t(f)(z) = Q_t(f)(z) = \frac{t}{\pi} \int_{-\infty}^{\infty} \frac{f(ze^{-iu})}{u^2+t^2} du$ or $U_t(f)(z) = P_t(f)(z) = \frac{1}{2t} \int_{-\infty}^{+\infty} f(ze^{-iu})e^{-|u|/t} du$ or $U_t(f)(z) = R_t(f)(z) = \frac{2t^3}{\pi} \int_{-\infty}^{+\infty} \frac{f(ze^{-iu})}{(u^2+t^2)^2} du$ or $U_t(f)(z) = W_t^*(f)(z) = \frac{1}{\sqrt{\pi t}} \int_{-\infty}^{+\infty} f(ze^{-iu})e^{-u^2/t} du$.

2) The complex forms are obtained simply replacing in the form of any $U_t(f)(x)$, the real variable $x \in \mathbb{R}$ by $z \in S$, where $S \subset \mathbb{C}$ is a strip, case when in fact we obtain some overconvergence results of these potentials.

Then, we will study the approximation properties of the complex Bessel-type potential, obtained from its real version by replacing the translation $x - y$ by the rotation ze^{-iy}, where $z = re^{ix} \in \mathbb{C}$, that is,

$$ B^\alpha(f)(z,t) = \frac{1}{\Gamma(\alpha/2)} \int_0^\infty \left[\int_{-\infty}^\infty \tau^{(\alpha/2)-1} e^{-\tau} W(y,\tau) f(ze^{-iy}, t-\tau) dy \right] d\tau. $$

Note that in order to exist $F_U^\alpha(f)(z)$ and $G_U^{\alpha,\beta}(f)(z)$ for all $|z| < R$, it is enough to suppose that the function $f(z)$ is analytic in $|z| < R$, with $R > 1$, while in order to exist $B^\alpha(f)(z,t)$, it is enough to suppose that the function $f(z,t)$ is in $L^p(\overline{\mathbb{D}_R} \times \mathbb{R})$, $1 \le p < \infty$, where $\mathbb{D}_R = \{z \in \mathbb{C}; |z| < R\}$. For the approximation properties of $B^\alpha(f)(z,t)$, we will suppose, in addition, that $f(z,t)$ is analytic in $\mathbb{D}_R$, $R > 1$, for any fixed $t \in \mathbb{R}$.

For $R > 0$ let us denote $\mathbb{D}_R = \{z \in \mathbb{C}; |z| < R\}$.

The first main result is the following.

Theorem 2.2.1 (Gal [56]). *Let us suppose that $\alpha > 0$ and that $f : \mathbb{D}_R \to \mathbb{C}$, with $R > 1$, is analytic in $\mathbb{D}_R$, that is, $f(z) = \sum_{k=0}^{\infty} a_k z^k$, for all $z \in \mathbb{D}_R$.*

(i) For $U_t(f)(z) = \frac{t}{\pi} \int_{-\infty}^{\infty} \frac{f(ze^{-iu})}{u^2+t^2} du$, we have that $F_U^\alpha(f)(z)$ is analytic in $\mathbb{D}_R$ and we can write

$$ F_U^\alpha(f)(z) = \sum_{k=0}^{\infty} a_k \cdot \frac{1}{(k+1)^\alpha} \cdot z^k, z \in \mathbb{D}_R. $$

Also, if f is not constant for $q = 0$ and not a polynomial of degree $\le q-1$ for $q \in \mathbb{N}$, then for all $1 \le r < r_1 < R$, $q \in \mathbb{N} \cup \{0\}$, $\alpha \in (0,1]$, we have

$$ \|[F_U^\alpha(f)]^{(q)} - f^{(q)}\|_r \sim \alpha, $$

where $\|f\|_r = \sup\{|f(z)|; |z| \le r\}$ and the constants in the equivalence depend only on f, q, r, and r_1.

(ii) For $U_t(f)(z) = \frac{1}{2t} \int_{-\infty}^{+\infty} f(ze^{-iu})e^{-|u|/t} du$, we have that $F_U^\alpha(f)(z)$ is analytic in $\mathbb{D}_R$ and we can write

$$ F_U^\alpha(f)(z) = \sum_{k=0}^{\infty} a_k b_{k,\alpha} z^k, z \in \mathbb{D}_R, $$

where $b_{k,\alpha} = \frac{1}{\Gamma(\alpha)} \int_0^\infty \frac{t^{\alpha-1}e^{-t}}{1+t^2k^2} dt.$

Also, if f is not constant for $q = 0$ and not a polynomial of degree $\leq q-1$ for $q \in \mathbb{N}$, then for all $1 \leq r < r_1 < R$, $q \in \mathbb{N} \cup \{0\}$, $\alpha \in (0,1]$, we have

$$\|[F_U^\alpha(f)]^{(q)} - f^{(q)}\|_r \sim \alpha,$$

where the constants in the equivalence depend only on f, q, r, and r_1.

(iii) For $U_t(f)(z) = \frac{2t^3}{\pi} \int_{-\infty}^{+\infty} \frac{f(ze^{-iu})}{(u^2+t^2)^2} du$, we have that $F_U^\alpha(f)(z)$ is analytic in $\mathbb{D}_R$ and we can write

$$F_U^\alpha(f)(z) = \sum_{k=0}^\infty a_k \frac{1}{(k+1)^{\alpha+1}} [k(\alpha+1)+1] \cdot z^k, z \in \mathbb{D}_R.$$

Also, there exists $\alpha_0 \in (0,1]$ (absolute constant) such that if f is not constant for $q = 0$ and not a polynomial of degree $\leq q - 1$ for $q \in \mathbb{N}$, then for all $1 \leq r < r_1 < R$, $q \in \mathbb{N} \cup \{0\}$, $\alpha \in (0, \alpha_0]$, we have

$$\|[F_U^\alpha(f)]^{(q)} - f^{(q)}\|_r \sim \alpha,$$

where the constants in the equivalence depend only on f, q, r, and r_1.

(iv) For $U_t(f)(z) = \frac{1}{\sqrt{\pi t}} \int_{-\infty}^{+\infty} f(ze^{-iu})e^{-u^2/t} du$, we have that $F_U^\alpha(f)(z)$ is analytic in $\mathbb{D}_R$ and we can write

$$F_U^\alpha(f)(z) = \sum_{k=0}^\infty a_k \cdot \frac{1}{(1+k^2/4)^{\alpha+1}} \cdot z^k, z \in \mathbb{D}_R.$$

Also, if f is not constant for $q = 0$ and not a polynomial of degree $\leq q-1$ for $q \in \mathbb{N}$, then for all $1 \leq r < r_1 < R$, $q \in \mathbb{N} \cup \{0\}$, $\alpha \in (0,1]$, we have

$$\|[F_U^\alpha(f)]^{(q)} - f^{(q)}\|_r \sim \alpha,$$

where the constants in the equivalence depend only on f, q, r, and r_1.

Proof. (i) By Gal [49], p. 213, Theorem 3.2.5, (i), $U_t(f)(z)$ is analytic (as function of z) in $\mathbb{D}_R$ and we can write

$$U_t(f)(z) = \sum_{k=0}^\infty a_k e^{-kt} z^k, \text{ for all } |z| < R \text{ and } t \geq 0.$$

Since $|\sum_{k=0}^\infty a_k e^{-kt} z^k| \leq \sum_{k=0}^\infty |a_k| \cdot |z|^k < \infty$, this implies that for fixed $|z| < R$, the series in t, $\sum_{k=0}^\infty a_k e^{-kt} z^k$ is uniformly convergent on $[0, \infty)$, and therefore we immediately can write

$$F_U^\alpha(f)(z) = \sum_{k=0}^\infty a_k z^k \frac{1}{\Gamma(\alpha)} \int_0^\infty t^{\alpha-1} e^{-(k+1)t} dt,$$

where by making use of the change of variable $(k+1)t = s$, we easily get that $\int_0^\infty t^{\alpha-1}e^{-(k+1)t}dt = \frac{\Gamma(\alpha)}{(k+1)^\alpha}$.

In other order of ideas, we easily can write

$$F_U^\alpha(f)(z) - f(z) = \frac{1}{\Gamma(\alpha)} \cdot \int_0^\infty t^{\alpha-1}e^{-t}[U_t(f)(z) - f(z)]dt,$$

which together with the estimate $|U_t(f)(z) - f(z)| \leq C_r(f)t$ in Gal [49], p. 213, Theorem 3.2.5, (iii), implies

$$|F_U^\alpha(f)(z) - f(z)| \leq \frac{1}{\Gamma(\alpha)} \cdot \int_0^\infty t^{\alpha-1}e^{-t}|U_t(f)(z) - f(z)|dt$$

$$\leq C_r(f)\frac{1}{\Gamma(\alpha)} \cdot \int_0^\infty t^\alpha e^{-t}dt = C_r(f) \cdot \frac{\Gamma(\alpha+1)}{\Gamma(\alpha)} = C_r(f)\alpha,$$

for all $|z| \leq r$, where $C_r(f) > 0$ is independent of z (and α) but depends on f and r.

Now, let $q \in \mathbb{N}\cup\{0\}$ and $1 \leq r < r_1 < R$. Denoting by γ the circle of radius r_1 and center 0, since for any $|z| \leq r$ and $v \in \gamma$, we have $|v - z| \geq r_1 - r$, by using Cauchy's formula, for all $|z| \leq r$ and $\alpha > 0$, we get

$$|[F_U^\alpha(f)]^{(q)}(z) - f^{(q)}(z)| = \frac{q!}{2\pi} \left| \int_\gamma \frac{F_U^\alpha(f)(z) - f(z)}{(v - z)^{q+1}}dv \right|$$

$$\leq C_{r_1}(f)\alpha \cdot \frac{q}{2\pi} \cdot \frac{2\pi r_1}{(r_1 - r)^{q+1}},$$

which proves the upper estimate

$$\|[F_U^\alpha(f)]^{(q)} - f^{(q)}\|_r \leq C^*\alpha,$$

with C^* depending only on f, q, r, and r_1.

It remains to prove the lower estimate. For this purpose, reasoning exactly as in the proof of Theorem 3.2.5, at pages 218–219 in the book of Gal [49], for $z = re^{i\varphi}$ and $p \in \mathbb{N} \cup \{0\}$, we get

$$\frac{1}{2\pi} \int_{-\pi}^\pi [f^{(q)}(z) - [U_t(f)]^{(q)}(z)]e^{-ip\varphi}d\varphi$$

$$= a_{q+p}(q + p)(q + p - 1)\ldots(p + 1)r^p[1 - e^{-(q+p)t}].$$

Multiplying above with $\frac{1}{\Gamma(\alpha)}t^{\alpha-1}e^{-t}$ and then integrating with respect to t, it follows

$$I := \frac{1}{\Gamma(\alpha)} \cdot \int_0^\infty \left\{ \frac{1}{2\pi} \int_{-\pi}^\pi [f^{(q)}(z) - [U_t(f)]^{(q)}(z)]e^{-ip\varphi} d\varphi \right\} t^{\alpha-1} e^{-t} dt$$

$$= a_{q+p}(q+p)(q+p-1)\ldots(p+1)r^p \frac{1}{\Gamma(\alpha)} \int_0^\infty t^{\alpha-1} e^{-t}[1 - e^{-(q+p)t}]dt$$

$$= a_{q+p}(q+p)(q+p-1)\ldots(p+1)r^p \left[1 - \frac{1}{(q+p+1)^\alpha} \right],$$

because taking into account that by making use of the change of variable $(q+p+1)t = s$, we easily get that

$$\frac{1}{\Gamma(\alpha)} \int_0^\infty t^{\alpha-1} e^{-t}[1 - e^{-(q+p)t}]dt = 1 - \frac{1}{\Gamma(\alpha)} \int_0^\infty t^{\alpha-1} e^{-(q+p+1)t} dt$$

$$= 1 - \frac{1}{(q+p+1)^\alpha}.$$

Applying Fubini's result to the double integral I and then passing to modulus, we easily obtain

$$\left| \frac{1}{2\pi} \int_{-\pi}^\pi e^{-ip\varphi} \left[\frac{1}{\Gamma(\alpha)} \int_0^\infty [f^{(q)}(z) - [U_t(f)]^{(q)}(z)] t^{\alpha-1} e^{-t} dt \right] d\varphi \right|$$

$$= |a_{q+p}|(q+p)(q+p-1)\ldots(p+1)r^p \left[1 - \frac{1}{(q+p+1)^\alpha} \right].$$

Since

$$\frac{1}{\Gamma(\alpha)} \int_0^\infty [f^{(q)}(z) - [U_t(f)]^{(q)}(z)] t^{\alpha-1} e^{-t} dt = f^{(q)}(z) - [F_U^\alpha(f)]^{(q)}(z),$$

the previous equality immediately implies

$$\left| \frac{1}{2\pi} \int_{-\pi}^\pi e^{-ip\varphi} \left[f^{(q)}(z) - (F_U^\alpha(f))^{(q)}(z) \right] d\varphi \right|$$

$$= |a_{q+p}|(q+p)(q+p-1)\ldots(p+1)r^p \left[1 - \frac{1}{(q+p+1)^\alpha} \right]$$

and

$$|a_{q+p}|(q+p)(q+p-1)\ldots(p+1)r^p \left[1 - \frac{1}{(q+p+1)^\alpha} \right] \le \|f^{(q)} - (F_U^\alpha(f))^{(q)}\|_r.$$

First take $q = 0$. From the previous inequality, we immediately obtain

$$|a_p|r^p \left(1 - \frac{1}{(p+1)^\alpha} \right) \le \|f - F_U^\alpha(f)\|_r.$$

In what follows, denoting $V_\alpha = \inf_{p \geq 1} \left(1 - \frac{1}{(p+1)^\alpha}\right)$, we clearly get $V_\alpha = 1 - \frac{1}{2^\alpha}$.

Denoting $g(x) = 2^{-x}$, by the mean value theorem, there exists $\xi \in (0, \alpha) \subset (0, 1]$ such that

$$V_\alpha = g(0) - g(\alpha) = -\alpha g'(\xi) = \alpha \cdot 2^{-\xi} ln(2) \geq \alpha 2^{-\alpha} ln(2) \geq \alpha 2^{-1} ln(2),$$

which immediately implies

$$\alpha \cdot \frac{ln(2)}{2} \cdot r^p \cdot |a_p| \leq \|f - F_U^\alpha(f)\|_r,$$

that is,

$$\frac{ln(2)}{2} \cdot r^p \cdot |a_p| \leq \frac{\|f - F_U^\alpha(f)\|_r}{\alpha},$$

for all $p \geq 1$ and $\alpha \in (0, 1]$.

This implies that if there exists a subsequence $(\alpha_k)_k$ in $(0, 1]$ with $\lim_{k \to \infty} \alpha_k = 0$ and such that $\lim_{k \to \infty} \frac{\|F_U^\alpha(f) - f\|_r}{\alpha_k} = 0$, then $a_p = 0$ for all $p \geq 1$, that is, f is constant on $\overline{\mathbb{D}}_r$.

Therefore, if f is not a constant function, then $\inf_{\alpha \in (0, 1]} \frac{\|F_U^\alpha(f) - f\|_r}{\alpha} > 0$, which implies that there exists a constant $C_r(f) > 0$ such that $\frac{\|F_U^\alpha(f) - f\|_r}{\alpha} \geq C_r(f)$, for all $\alpha \in (0, 1]$, that is,

$$\|F_U^\alpha(f) - f\|_r \geq C_r(f)\alpha, \text{ for all } \alpha \in (0, 1].$$

Now, consider $q \geq 1$ and denote $V_{q,\alpha} = \inf_{p \geq 0}(1 - \frac{1}{(q+p+1)^\alpha})$. Evidently that we have $V_{q,\alpha} \geq \inf_{p \geq 1}(1 - \frac{1}{(p+1)^\alpha}) \geq \alpha \cdot \frac{ln(2)}{2}$.

Reasoning as in the case of $q = 0$ we obtain

$$\frac{\|[F_U^\alpha(f)]^{(q)} - f^{(q)}\|_r}{\alpha} \geq |a_{q+p}| \frac{(q+p)!}{p!} \cdot \frac{ln(2)}{2} \cdot r^p,$$

for all $p \geq 0$ and $\alpha \in (0, 1]$.

This implies that if there exists a subsequence $(\alpha_k)_k$ in $(0, 1]$ with $\lim_{k \to \infty} \alpha_k = 0$ and such that $\lim_{k \to \infty} \frac{\|[F_U^\alpha(f)]^{(q)} - f^{(q)}\|_r}{\alpha_k} = 0$, then $a_{q+p} = 0$ for all $p \geq 0$, that is, f is a polynomial of degree $\leq q - 1$ on $\overline{\mathbb{D}}_r$.

Therefore, because by hypothesis f is not a polynomial of degree $\leq q - 1$, we obtain $\inf_{\alpha \in (0, 1]} \frac{\|[F_U^\alpha(f)]^{(q)} - f^{(q)}\|_r}{\alpha} > 0$, which implies that there exists a constant $C_{r,q}(f) > 0$ such that $\frac{\|[F_U^\alpha(f)]^{(q)} - f^{(q)}\|_r}{\alpha} \geq C_{r,q}(f)$, for all $\alpha \in (0, 1]$, that is,

$$\|[F_U^\alpha(f)]^{(q)} - f^{(q)}\|_r \geq C_{r,q}(f)\alpha, \text{ for all } \alpha \in (0, 1].$$

(ii) By Gal [49], p. 206, Theorem 3.2.1, (i), $U_t(f)(z)$ is analytic (as function of z) in $\mathbb{D}_R$ and we can write

$$U_t(f)(z) = \sum_{k=0}^{\infty} \frac{a_k}{1+t^2k^2} z^k, \text{ for all } |z| < R \text{ and } t \geq 0.$$

Since $\left| \sum_{k=0}^{\infty} \frac{a_k}{1+t^2k^2} z^k \right| \leq \sum_{k=0}^{\infty} |a_k| \cdot |z|^k < \infty$, this implies that for fixed $|z| < R$, the series in t, $\sum_{k=0}^{\infty} \frac{a_k}{1+t^2k^2} z^k$ is uniformly convergent on $[0, \infty)$, and therefore we immediately can write

$$F_U^\alpha(f)(z) = \sum_{k=0}^{\infty} a_k z^k \frac{1}{\Gamma(\alpha)} \int_0^\infty \frac{t^{\alpha-1}e^{-t}}{1+t^2k^2} dt.$$

In other order of ideas, we easily can write

$$F_U^\alpha(f)(z) - f(z) = \frac{1}{\Gamma(\alpha)} \cdot \int_0^\infty t^{\alpha-1}e^{-t}[U_t(f)(z) - f(z)]dt,$$

which together with the estimate $|U_t(f)(z) - f(z)| \leq C_r(f)t^2$ in Gal [49], p. 207, Theorem 3.2.1, (iv), implies

$$|F_U^\alpha(f)(z) - f(z)| \leq \frac{1}{\Gamma(\alpha)} \cdot \int_0^\infty t^{\alpha-1}e^{-t}|U_t(f)(z) - f(z)|dt$$

$$\leq C_r(f)\frac{1}{\Gamma(\alpha)} \cdot \int_0^\infty t^{\alpha+1}e^{-t}dt = C_r(f) \cdot \frac{\Gamma(\alpha+2)}{\Gamma(\alpha)} = C_r(f)\alpha(\alpha+1) \leq 2C_r(f)\alpha,$$

for all $|z| \leq r$, where $C_r(f) > 0$ is independent of z (and α) but depends on f and r.

Now, let $q \in \mathbb{N} \cup \{0\}$ and $1 \leq r < r_1 < R$. Denoting by γ the circle of radius r_1 and center 0, since for any $|z| \leq r$ and $v \in \gamma$, we have $|v - z| \geq r_1 - r$, by using Cauchy's formula, for all $|z| \leq r$ and $\alpha > 0$, we get

$$|[F_U^\alpha(f)]^{(q)}(z) - f^{(q)}(z)| = \frac{q!}{2\pi} \left| \int_\gamma \frac{F_U^\alpha(f)(z) - f(z)}{(v-z)^{q+1}} dv \right|$$

$$\leq 2C_{r_1}(f)\alpha \cdot \frac{q}{2\pi} \cdot \frac{2\pi r_1}{(r_1 - r)^{q+1}},$$

which proves the upper estimate

$$\|[F_U^\alpha(f)]^{(q)} - f^{(q)}\|_r \leq C^*\alpha,$$

with C^* depending only on f, q, r, and r_1.

It remains to prove the lower estimate. For this purpose, reasoning exactly as in the proof of Theorem 3.2.1, at pages 209–210 in the book of Gal [49], for $z = re^{i\varphi}$ and $p \in \mathbb{N} \cup \{0\}$, we get

$$\frac{1}{2\pi}\int_{-\pi}^{\pi}[f^{(q)}(z)-[U_t(f)]^{(q)}(z)]e^{-ip\varphi}d\varphi$$

$$=a_{q+p}(q+p)(q+p-1)\ldots(p+1)r^p\cdot\frac{t^2(q+p)^2}{1+t^2(q+p)^2}.$$

Multiplying above with $\frac{1}{\Gamma(\alpha)}t^{\alpha-1}e^{-t}$ and then integrating with respect to t, it follows

$$I:=\frac{1}{\Gamma(\alpha)}\cdot\int_0^\infty\left\{\frac{1}{2\pi}\int_{-\pi}^{\pi}[f^{(q)}(z)-[U_t(f)]^{(q)}(z)]e^{-ip\varphi}d\varphi\right\}t^{\alpha-1}e^{-t}dt$$

$$=a_{q+p}(q+p)(q+p-1)\ldots(p+1)r^p\frac{1}{\Gamma(\alpha)}\int_0^\infty t^{\alpha-1}e^{-t}\left[\frac{t^2(q+p)^2}{1+t^2(q+p)^2}\right]dt.$$

Applying Fubini's result to the double integral I and then passing to modulus, we easily obtain

$$\left|\frac{1}{2\pi}\int_{-\pi}^{\pi}e^{-ip\varphi}\left[\frac{1}{\Gamma(\alpha)}\int_0^\infty[f^{(q)}(z)-[U_t(f)]^{(q)}(z)]t^{\alpha-1}e^{-t}dt\right]d\varphi\right|$$

$$=|a_{q+p}|(q+p)(q+p-1)\ldots(p+1)r^p\left[\frac{1}{\Gamma(\alpha)}\int_0^\infty t^{\alpha-1}e^{-t}\left[\frac{t^2(q+p)^2}{1+t^2(q+p)^2}\right]dt\right].$$

Since

$$\frac{1}{\Gamma(\alpha)}\int_0^\infty[f^{(q)}(z)-[U_t(f)]^{(q)}(z)]t^{\alpha-1}e^{-t}dt=f^{(q)}(z)-[F_U^\alpha(f)]^{(q)}(z),$$

the previous equality immediately implies

$$\left|\frac{1}{2\pi}\int_{-\pi}^{\pi}e^{-ip\varphi}\left[f^{(q)}(z)-(F_U^\alpha(f))^{(q)}(z)\right]d\varphi\right|$$

$$=|a_{q+p}|(q+p)(q+p-1)\ldots(p+1)r^p\left[\frac{1}{\Gamma(\alpha)}\int_0^\infty t^{\alpha-1}e^{-t}\left[\frac{t^2(q+p)^2}{1+t^2(q+p)^2}\right]dt\right]$$

and

$$|a_{q+p}|(q+p)(q+p-1)\ldots(p+1)r^p\left[\frac{1}{\Gamma(\alpha)}\int_0^\infty t^{\alpha-1}e^{-t}\left[\frac{t^2(q+p)^2}{1+t^2(q+p)^2}\right]dt\right]$$

$$\leq\|f^{(q)}-(F_U^\alpha(f))^{(q)}\|_r.$$

First take $q=0$. From the previous inequality we immediately obtain

$$|a_p|r^p\left(\frac{1}{\Gamma(\alpha)}\int_0^\infty t^{\alpha-1}e^{-t}\left[\frac{t^2p^2}{1+t^2p^2}\right]dt\right)\leq\|f-F_U^\alpha(f)\|_r.$$

In what follows, denoting $V_\alpha = \inf_{p \geq 1} \left(\frac{1}{\Gamma(\alpha)} \int_0^\infty t^{\alpha-1} e^{-t} \left[\frac{t^2 p^2}{1+t^2 p^2} \right] dt \right)$, we clearly get

$$V_\alpha = \frac{1}{\Gamma(\alpha)} \int_0^\infty t^{\alpha-1} e^{-t} \left[\frac{t^2}{1+t^2} \right] dt.$$

Taking into account that $1 + t^2 \leq 2e^t$ for all $t \geq 0$, we obtain

$$V_\alpha \geq \frac{1}{\Gamma(\alpha)} \int_0^\infty t^{\alpha+1} e^{-2t} dt = \frac{\Gamma(\alpha+2)}{2^{\alpha+2}\Gamma(\alpha)} = \frac{\alpha}{4} \cdot \frac{\alpha+1}{2^\alpha} \geq C\alpha,$$

since the function $f(x) = \frac{x+1}{2^x}$ is strictly positive and continuous in $[0, 1]$.
This immediately implies

$$C \cdot r^p \cdot |a_p| \leq \frac{\|f - F_U^\alpha(f)\|_r}{\alpha},$$

for all $p \geq 1$ and $\alpha \in (0, 1]$.

Now, if a subsequence $(\alpha_k)_k$ in $(0, 1]$ with $\lim_{k \to \infty} \alpha_k = 0$ would exist and such that $\lim_{k \to \infty} \frac{\|F_U^\alpha(f)-f\|_r}{\alpha_k} = 0$, then $a_p = 0$ for all $p \geq 1$, that is, f would be constant on $\overline{\mathbb{D}}_r$. Therefore, if f is not a constant function, then $\inf_{\alpha \in (0,1]} \frac{\|F_U^\alpha(f)-f\|_r}{\alpha} > 0$, which implies that there exists a constant $C_r(f) > 0$ such that $\frac{\|F_U^\alpha(f)-f\|_r}{\alpha} \geq C_r(f)$, for all $\alpha \in (0, 1]$, that is,

$$\|F_U^\alpha(f) - f\|_r \geq C_r(f)\alpha, \text{ for all } \alpha \in (0, 1].$$

Now, consider $q \geq 1$ and denote

$$V_{q,\alpha} = \inf_{p \geq 0} \left(\frac{1}{\Gamma(\alpha)} \int_0^\infty t^{\alpha-1} e^{-t} \left[\frac{t^2(q+p)^2}{1+t^2(q+p)^2} \right] \right).$$

Evidently that we have $V_{q,\alpha} \geq \inf_{p \geq 1} \left(\frac{1}{\Gamma(\alpha)} \int_0^\infty t^{\alpha-1} e^{-t} \left[\frac{t^2 p^2}{1+t^2 p^2} \right] dt \right) \geq \alpha \cdot C$.
Reasoning as in the case of $q = 0$, we obtain

$$\frac{\|[F_U^\alpha(f)]^{(q)} - f^{(q)}\|_r}{\alpha} \geq |a_{q+p}| \frac{(q+p)!}{p!} \cdot C \cdot r^p,$$

for all $p \geq 0$ and $\alpha \in (0, 1]$.

This implies that if there exists a subsequence $(\alpha_k)_k$ in $(0, 1]$ with $\lim_{k \to \infty} \alpha_k = 0$ and such that $\lim_{k \to \infty} \frac{\|[F_U^\alpha(f)]^{(q)}-f^{(q)}\|_r}{\alpha_k} = 0$, then $a_{q+p} = 0$ for all $p \geq 0$, that is, f is a polynomial of degree $\leq q - 1$ on $\overline{\mathbb{D}}_r$.

Therefore, because by hypothesis f is not a polynomial of degree $\leq q - 1$, we obtain $\inf_{\alpha \in (0,1]} \frac{\|[F_U^\alpha(f)]^{(q)}-f^{(q)}\|_r}{\alpha} > 0$, which implies that there exists a constant $C_{r,q}(f) > 0$ such that $\frac{\|[F_U^\alpha(f)]^{(q)}-f^{(q)}\|_r}{\alpha} \geq C_{r,q}(f)$, for all $\alpha \in (0, 1]$, that is,

$$\|[F_U^\alpha(f)]^{(q)} - f^{(q)}\|_r \geq C_{r,q}(f)\alpha, \text{ for all } \alpha \in (0,1].$$

(iii) By Gal [49], p. 213, Theorem 3.2.5, (i), $U_t(f)(z)$ is analytic (as function of z) in $\mathbb{D}_R$ and we can write

$$U_t(f)(z) = \sum_{k=0}^{\infty} a_k(1 + kt)e^{-kt}z^k, \text{ for all } |z| < R \text{ and } t \geq 0.$$

Since $|\sum_{k=0}^{\infty} a_k e^{-kt}(1+kt)z^k| \leq 2\sum_{k=0}^{\infty} |a_k| \cdot |z|^k < \infty$, this implies that for fixed $|z| < R$, the series in t, $\sum_{k=0}^{\infty} a_k(1 + kt)e^{-kt}z^k$ is uniformly convergent on $[0, \infty)$, and therefore we immediately can write

$$F_U^\alpha(f)(z) = \sum_{k=0}^{\infty} a_k z^k \frac{1}{\Gamma(\alpha)} \int_0^\infty t^{\alpha-1}(1 + kt)e^{-(k+1)t}dt,$$

where by making use of the change of variable $(k+1)t = s$, we easily get that $\int_0^\infty t^{\alpha-1}e^{-(k+1)t}dt = \frac{\Gamma(\alpha)}{(k+1)^\alpha}$ and therefore we immediately obtain

$$F_U^\alpha(f)(z) = \sum_{k=0}^{\infty} a_k \frac{1}{(k+1)^{\alpha+1}}[k(\alpha+1) + 1]z^k.$$

In other order of ideas, we easily can write

$$F_U^\alpha(f)(z) - f(z) = \frac{1}{\Gamma(\alpha)} \cdot \int_0^\infty t^{\alpha-1}e^{-t}[U_t(f)(z) - f(z)]dt,$$

which together with the estimate $|U_t(f)(z) - f(z)| \leq C_r(f)t^2$ in Gal [49], p. 213–214, Theorem 3.2.5, (iv), implies

$$|F_U^\alpha(f)(z) - f(z)| \leq \frac{1}{\Gamma(\alpha)} \cdot \int_0^\infty t^{\alpha-1}e^{-t}|U_t(f)(z) - f(z)|dt$$

$$\leq C_r(f)\frac{1}{\Gamma(\alpha)} \cdot \int_0^\infty t^{\alpha+1}e^{-t}dt = C_r(f) \cdot \frac{\Gamma(\alpha+2)}{\Gamma(\alpha)} = C_r(f)\alpha(\alpha+1) \leq 2C_r(f)\alpha,$$

for all $|z| \leq r$, where $C_r(f) > 0$ is independent of z (and α) but depends on f and r.

Now, let $q \in \mathbb{N} \cup \{0\}$ and $1 \leq r < r_1 < R$. Denoting by γ the circle of radius r_1 and center 0, since for any $|z| \leq r$ and $v \in \gamma$, we have $|v - z| \geq r_1 - r$, by using Cauchy's formula, for all $|z| \leq r$ and $\alpha > 0$, we get

$$|[F_U^\alpha(f)]^{(q)}(z) - f^{(q)}(z)| = \frac{q!}{2\pi}\left|\int_\gamma \frac{F_U^\alpha(f)(z) - f(z)}{(v - z)^{q+1}}dv\right|$$

$$\leq C_{r_1}(f)\alpha \cdot \frac{q}{2\pi} \cdot \frac{2\pi r_1}{(r_1 - r)^{q+1}},$$

which proves the upper estimate

$$\|[F_U^\alpha(f)]^{(q)} - f^{(q)}\|_r \le C^*\alpha,$$

with C^* depending only on f, q, r, and r_1.

It remains to prove the lower estimate. For this purpose, reasoning exactly as in the proof of Theorem 3.2.5, at pages 219–220 in the book of Gal [49], for $z = re^{i\varphi}$ and $p \in \mathbb{N} \cup \{0\}$, we get

$$\frac{1}{2\pi} \int_{-\pi}^{\pi} [f^{(q)}(z) - [U_t(f)]^{(q)}(z)]e^{-ip\varphi} d\varphi$$

$$= a_{q+p}(q+p)(q+p-1)\ldots(p+1)r^p[1 - (1+(q+p)t)e^{-(q+p)t}].$$

Multiplying above with $\frac{1}{\Gamma(\alpha)}t^{\alpha-1}e^{-t}$ and then integrating with respect to t, it follows

$$I := \frac{1}{\Gamma(\alpha)} \cdot \int_0^\infty \left\{ \frac{1}{2\pi} \int_{-\pi}^{\pi} [f^{(q)}(z) - [U_t(f)]^{(q)}(z)]e^{-ip\varphi} d\varphi \right\} t^{\alpha-1}e^{-t} dt$$

$$= a_{q+p}(q+p)(q+p-1)\ldots(p+1)r^p$$

$$\cdot \frac{1}{\Gamma(\alpha)} \int_0^\infty t^{\alpha-1}e^{-t} \left[1 - (1+(q+p)t)e^{-(q+p)t}\right] dt.$$

Applying Fubini's result to the double integral I and then passing to modulus, we easily obtain

$$\left| \frac{1}{2\pi} \int_{-\pi}^{\pi} e^{-ip\varphi} \left[\frac{1}{\Gamma(\alpha)} \int_0^\infty [f^{(q)}(z) - [U_t(f)]^{(q)}(z)]t^{\alpha-1}e^{-t} dt \right] d\varphi \right|$$

$$= |a_{q+p}|(q+p)(q+p-1)\ldots(p+1)r^p$$

$$\cdot \left[\frac{1}{\Gamma(\alpha)} \int_0^\infty t^{\alpha-1}e^{-t} \left[1 - (1+(q+p)t)e^{-(q+p)t}\right] dt \right].$$

Since

$$\frac{1}{\Gamma(\alpha)} \int_0^\infty [f^{(q)}(z) - [U_t(f)]^{(q)}(z)]t^{\alpha-1}e^{-t} dt = f^{(q)}(z) - [F_U^\alpha(f)]^{(q)}(z),$$

the previous equality immediately implies

$$\left| \frac{1}{2\pi} \int_{-\pi}^{\pi} e^{-ip\varphi} \left[f^{(q)}(z) - (F_U^\alpha(f))^{(q)}(z) \right] d\varphi \right|$$

$$= |a_{q+p}|(q+p)(q+p-1)\ldots(p+1)r^p$$

$$\cdot \left[\frac{1}{\Gamma(\alpha)} \int_0^\infty t^{\alpha-1}e^{-t} \left[1 - (1+(q+p)t)e^{-(q+p)t}\right] dt \right]$$

and

$$|a_{q+p}|(q+p)(q+p-1)\ldots(p+1)r^p$$

$$\cdot \left[\frac{1}{\Gamma(\alpha)} \int_0^\infty t^{\alpha-1}e^{-t} \left[1 - (1+(q+p)t)e^{-(q+p)t}\right] dt \right] \leq \|f^{(q)} - (F_U^\alpha(f))^{(q)}\|_r.$$

First take $q = 0$. From the previous inequality, we immediately obtain

$$|a_p|r^p \left(\frac{1}{\Gamma(\alpha)} \int_0^\infty t^{\alpha-1}e^{-t} \left[1 - (1+pt)e^{-pt}\right] dt \right) \leq \|f - F_U^\alpha(f)\|_r.$$

In what follows, denoting

$$V_\alpha = \inf_{p \geq 1} \left(\frac{1}{\Gamma(\alpha)} \int_0^\infty t^{\alpha-1}e^{-t} \left[1 - (1+pt)e^{-pt}\right] dt \right),$$

by simple calculation, we get

$$V_\alpha = \frac{1}{\Gamma(\alpha)} \int_0^\infty t^{\alpha-1}e^{-t} \left[1 - (1+t)e^{-t}\right] dt = 1 - \frac{1}{2^\alpha} - \frac{\alpha}{2^{\alpha+1}}.$$

But there exists $\alpha_0 \in (0,1]$, such that if C is an absolute constant with $0 < C < ln(2) - \frac{1}{2}$, then we have

$$1 - \frac{1}{2^\alpha} - \frac{\alpha}{2^{\alpha+1}} \geq C\alpha, \text{ for all } \alpha \in [0, \alpha_0].$$

Indeed, denoting $g(\alpha) = 1 - \frac{1}{2^\alpha} - \frac{\alpha}{2^{\alpha+1}} - C\alpha$, we have $g(0) = 0$ and $g'(\alpha) = 2^{-\alpha}ln(2) - \frac{1}{2^{\alpha+1}} + \frac{\alpha(\alpha+1)ln(2)}{2^{\alpha+1}} - C$, which implies $g'(0) = ln(2) - \frac{1}{2} - C > 0$. Since $g'(\alpha)$ obviously is continuous with respect to α, there exists $\alpha_0 > 0$ such that $g'(\alpha) > 0$ for all $\alpha \in [0, \alpha_0]$, that is, $V_\alpha \geq C\alpha$, for all $\alpha \in [0, \alpha_0]$.

This immediately implies

$$C \cdot r^p \cdot |a_p| \leq \frac{\|f - F_U^\alpha(f)\|_r}{\alpha},$$

for all $p \geq 1$ and $\alpha \in (0, \alpha_0]$.

Now, if a subsequence $(\alpha_k)_k$ in $(0, \alpha_0]$ with $\lim_{k\to\infty} \alpha_k = 0$ would exist and such that $\lim_{k\to\infty} \frac{\|F_U^\alpha(f)-f\|_r}{\alpha_k} = 0$, then $a_p = 0$ for all $p \geq 1$, that is, f would be constant on $\overline{\mathbb{D}}_r$. Therefore, if f is not a constant function, then $\inf_{\alpha \in (0,\alpha_0]} \frac{\|F_U^\alpha(f)-f\|_r}{\alpha} > 0$, which implies that there exists a constant $C_r(f) > 0$ such that $\frac{\|F_U^\alpha(f)-f\|_r}{\alpha} \geq C_r(f)$, for all $\alpha \in (0, \alpha_0]$, that is,

$$\|F_U^\alpha(f) - f\|_r \geq C_r(f)\alpha, \text{ for all } \alpha \in (0, \alpha_0].$$

Now, consider $q \geq 1$ and denote

$$V_{q,\alpha} = \inf_{p \geq 0} \left(\frac{1}{\Gamma(\alpha)} \int_0^\infty t^{\alpha-1} e^{-t} \left[1 - (1 + (q+p)t)e^{-(q+p)t} \right] \right).$$

Evidently that we have $V_{q,\alpha} \geq \inf_{p \geq 1} \left(\frac{1}{\Gamma(\alpha)} \int_0^\infty t^{\alpha-1} e^{-t} \left[1 - (1+pt)e^{-pt} \right] dt \right) \geq \alpha \cdot C$, for $C \in (0, \ln(2) - 1/2)$ and $\alpha \in [0, \alpha_0]$.

Reasoning exactly as in the case of $q = 0$ and as in the previous case (ii), we easily obtain that because by hypothesis f is not a polynomial of degree $\leq q - 1$, there exists a constant $C_{r,q}(f) > 0$ such that

$$\| [F_U^\alpha(f)]^{(q)} - f^{(q)} \|_r \geq C_{r,q}(f)\alpha, \text{ for all } \alpha \in (0, \alpha_0].$$

(iv) By Gal [49], p. 223, Theorem 3.2.8, (i), $U_t(f)(z)$ is analytic (as function of z) in $\mathbb{D}_R$ and we can write

$$U_t(f)(z) = \sum_{k=0}^\infty a_k e^{-k^2 t/4} z^k, \text{ for all } |z| < R \text{ and } t \geq 0.$$

Since $|\sum_{k=0}^\infty a_k e^{-k^2 t/4} z^k| \leq \sum_{k=0}^\infty |a_k| \cdot |z|^k < \infty$, this implies that for fixed $|z| < R$, the series in t, $\sum_{k=0}^\infty a_k e^{-k^2 t/4} z^k$ is uniformly convergent on $[0, \infty)$, and therefore we immediately can write

$$F_U^\alpha(f)(z) = \sum_{k=0}^\infty a_k z^k \frac{1}{\Gamma(\alpha)} \int_0^\infty t^{\alpha-1} e^{-(1+k^2/4)t} dt,$$

where by making use of the change of variable $(1 + k^2/4)t = s$, we easily get that $\int_0^\infty t^{\alpha-1} e^{-(1+k^2/4)t} dt = \frac{\Gamma(\alpha)}{(1+k^2/4)^\alpha}$ and therefore we immediately obtain

$$F_U^\alpha(f)(z) = \sum_{k=0}^\infty a_k \frac{1}{(1 + k^2/4)^{\alpha+1}} z^k.$$

In other order of ideas, we easily can write

$$F_U^\alpha(f)(z) - f(z) = \frac{1}{\Gamma(\alpha)} \cdot \int_0^\infty t^{\alpha-1} e^{-t} [U_t(f)(z) - f(z)] dt,$$

which together with the estimate $|U_t(f)(z) - f(z)| \leq C_r(f)t$ in Gal [49], p. 224, Theorem 3.2.8, (iv), implies

$$|F_U^\alpha(f)(z) - f(z)| \leq \frac{1}{\Gamma(\alpha)} \cdot \int_0^\infty t^{\alpha-1} e^{-t} |U_t(f)(z) - f(z)| dt$$

$$\leq C_r(f) \frac{1}{\Gamma(\alpha)} \cdot \int_0^\infty t^\alpha e^{-t} dt = C_r(f) \cdot \frac{\Gamma(\alpha+1)}{\Gamma(\alpha)} = C_r(f)\alpha,$$

for all $|z| \leq r$, where $C_r(f) > 0$ is independent of z (and α) but depends on f and r.

Now, let $q \in \mathbb{N} \cup \{0\}$ and $1 \leq r < r_1 < R$. Denoting by γ the circle of radius r_1 and center 0, since for any $|z| \leq r$ and $v \in \gamma$, we have $|v - z| \geq r_1 - r$, by using Cauchy's formula, for all $|z| \leq r$ and $\alpha > 0$, we get

$$\left| [F_U^\alpha(f)]^{(q)}(z) - f^{(q)}(z) \right| = \frac{q!}{2\pi} \left| \int_\gamma \frac{F_U^\alpha(f)(z) - f(z)}{(v-z)^{q+1}} dv \right|$$

$$\leq C_{r_1}(f)\alpha \cdot \frac{q}{2\pi} \cdot \frac{2\pi r_1}{(r_1 - r)^{q+1}},$$

which proves the upper estimate

$$\| [F_U^\alpha(f)]^{(q)} - f^{(q)} \|_r \leq C^* \alpha,$$

with C^* depending only on f, q, r, and r_1.

It remains to prove the lower estimate. For this purpose, reasoning exactly as in the proof of Theorem 3.2.8, at pages 227–228 in the book of Gal [49], for $z = re^{i\varphi}$ and $p \in \mathbb{N} \cup \{0\}$, we get

$$\frac{1}{2\pi} \int_{-\pi}^{\pi} [f^{(q)}(z) - [U_t(f)]^{(q)}(z)] e^{-ip\varphi} d\varphi$$

$$= a_{q+p}(q+p)(q+p-1)\ldots(p+1)r^p [1 - e^{-(q+p)^2 t/4}].$$

Multiplying above with $\frac{1}{\Gamma(\alpha)} t^{\alpha-1} e^{-t}$ and then integrating with respect to t, it follows

$$I := \frac{1}{\Gamma(\alpha)} \cdot \int_0^\infty \left\{ \frac{1}{2\pi} \int_{-\pi}^{\pi} [f^{(q)}(z) - [U_t(f)]^{(q)}(z)] e^{-ip\varphi} d\varphi \right\} t^{\alpha-1} e^{-t} dt$$

$$= a_{q+p}(q+p)(q+p-1)\ldots(p+1)r^p$$

$$\cdot \frac{1}{\Gamma(\alpha)} \int_0^\infty t^{\alpha-1} e^{-t} \left[1 - e^{-(q+p)^2 t/4} \right] dt.$$

Applying Fubini's result to the double integral I and then passing to modulus, we easily obtain

$$\left| \frac{1}{2\pi} \int_{-\pi}^{\pi} e^{-ip\varphi} \left[\frac{1}{\Gamma(\alpha)} \int_0^\infty [f^{(q)}(z) - [U_t(f)]^{(q)}(z)] t^{\alpha-1} e^{-t} dt \right] d\varphi \right|$$

$$= |a_{q+p}|(q+p)(q+p-1)\ldots(p+1)r^p$$

$$\cdot \left[\frac{1}{\Gamma(\alpha)} \int_0^\infty t^{\alpha-1} e^{-t} \left[1 - e^{-(q+p)^2 t/4} \right] dt \right].$$

Since

$$\frac{1}{\Gamma(\alpha)} \int_0^\infty [f^{(q)}(z) - [U_t(f)]^{(q)}(z)] t^{\alpha-1} e^{-t} dt = f^{(q)}(z) - [F_U^\alpha(f)]^{(q)}(z),$$

the previous equality immediately implies

$$\left| \frac{1}{2\pi} \int_{-\pi}^\pi e^{-ip\varphi} \left[f^{(q)}(z) - (F_U^\alpha(f))^{(q)}(z) \right] d\varphi \right|$$

$$= |a_{q+p}|(q+p)(q+p-1)\dots(p+1)r^p$$

$$\cdot \left[\frac{1}{\Gamma(\alpha)} \int_0^\infty t^{\alpha-1} e^{-t} \left[1 - e^{-(q+p)^2 t/4} \right] dt \right]$$

and

$$|a_{q+p}|(q+p)(q+p-1)\dots(p+1)r^p$$

$$\cdot \left[\frac{1}{\Gamma(\alpha)} \int_0^\infty t^{\alpha-1} e^{-t} \left[1 - e^{-(q+p)^2 t/4} \right] dt \right] \le \| f^{(q)} - (F_U^\alpha(f))^{(q)} \|_r.$$

First take $q = 0$. From the previous inequality, we immediately obtain

$$|a_p| r^p \left(\frac{1}{\Gamma(\alpha)} \int_0^\infty t^{\alpha-1} e^{-t} \left[1 - e^{-p^2 t/4} \right] dt \right) \le \| f - F_U^\alpha(f) \|_r.$$

In what follows, denoting

$$V_\alpha = \inf_{p \ge 1} \left(\frac{1}{\Gamma(\alpha)} \int_0^\infty t^{\alpha-1} e^{-t} \left[1 - e^{-p^2 t/4} \right] dt \right),$$

by simple calculation, we get

$$V_\alpha = \frac{1}{\Gamma(\alpha)} \int_0^\infty t^{\alpha-1} e^{-t} \left[1 - e^{-t/4} \right] dt = 1 - \left(\frac{4}{5} \right)^\alpha.$$

Denoting $g(x) = \left(\frac{4}{5} \right)^x$, by the mean value theorem, there exists $\xi \in (0, \alpha) \subset (0, 1]$ such that

$$V_\alpha = g(0) - g(\alpha) = -\alpha g'(\xi) = \alpha \cdot \left(\frac{4}{5} \right)^\xi \ln \left(\frac{4}{5} \right) \ge \alpha \left(\frac{4}{5} \right)^\alpha \ln \left(\frac{4}{5} \right)$$

$$\ge \alpha \left(\frac{4}{5} \right) \ln \left(\frac{4}{5} \right),$$

which immediately implies

$$\left(\frac{4}{5} \right) \ln \left(\frac{4}{5} \right) \cdot r^p \cdot |a_p| \le \frac{\| f - F_U^\alpha(f) \|_r}{\alpha},$$

for all $p \ge 1$ and $\alpha \in (0, 1]$.

Reasoning now exactly as in the proof of the above point (i), we similarly get that if f is not a constant function, then there exists a constant $C_r(f) > 0$ such that

$$\|F_U^\alpha(f) - f\|_r \geq C_r(f)\alpha, \text{ for all } \alpha \in (0, 1].$$

Now, consider $q \geq 1$ and denote

$$V_{q,\alpha} = \inf_{p \geq 0} \left(\frac{1}{\Gamma(\alpha)} \int_0^\infty t^{\alpha-1} e^{-t} \left[1 - e^{-(q+p)^2 t/4} \right] \right).$$

Evidently that we have

$$V_{q,\alpha} \geq \inf_{p \geq 1} \left(\frac{1}{\Gamma(\alpha)} \int_0^\infty t^{\alpha-1} e^{-t} \left[1 - e^{-p^2 t/4} \right] dt \right) \geq \alpha \cdot C,$$

for all $\alpha \in [0, 1]$.

Reasoning in continuation exactly as in the case of $q = 0$ and as in the previous case (i), we easily obtain that because by hypothesis f is not a polynomial of degree $\leq q - 1$, there exists a constant $C_{r,q}(f) > 0$ such that

$$\|[F_U^\alpha(f)]^{(q)} - f^{(q)}\|_r \geq C_{r,q}(f)\alpha, \text{ for all } \alpha \in (0, 1].$$

The theorem is proved. $\square$

Replacing now everywhere in Theorem 2.2.1 the $\Gamma(\alpha)$ function by the $Beta(\alpha, \beta)$ function and considering the construction of $G_U^{\alpha,\beta}(f)(z)$ defined in the direction 1) just before the statement of Theorem 2.4.1, we have the following result.

Theorem 2.2.2 (Gal [57]). *Let us suppose that $0 < \alpha \leq \beta \leq 1$, $\alpha + \beta \geq 1$ and that $f : \mathbb{D}_R \to \mathbb{C}$, with $R > 1$, is analytic in $\mathbb{D}_R$, that is, $f(z) = \sum_{k=0}^\infty a_k z^k$, for all $z \in \mathbb{D}_R$.*

(i) For $U_t(f)(z) = \frac{t}{\pi} \int_{-\infty}^\infty \frac{f(ze^{-iu})}{u^2+t^2} du$, we have that $G_U^{\alpha,\beta}(f)(z)$ is analytic in $\mathbb{D}_R$ and we can write

$$G_U^{\alpha,\beta}(f)(z) = \sum_{k=0}^\infty a_k b_k(\alpha, \beta) \cdot z^k, z \in \mathbb{D}_R,$$

where

$$b_k(\alpha, \beta) = \frac{1}{Beta(\alpha, \beta)} \int_0^1 t^{\alpha-1}(1 - t)^{\beta-1} e^{-kt} dt.$$

Also, if f is not constant for $q = 0$ and not a polynomial of degree $\leq q-1$ for $q \in \mathbb{N}$, then for all $1 \leq r < r_1 < R$, $q \in \mathbb{N} \cup \{0\}$, $\alpha \in (0, \beta]$, we have

$$\|[G_U^{\alpha,\beta}(f)]^{(q)} - f^{(q)}\|_r \sim \alpha,$$

where $\|f\|_r = \sup\{|f(z)|; |z| \leq r\}$ and the constants in the equivalence depend only on f, q, r, r_1, β.

(ii) For $U_t(f)(z) = \frac{1}{2t} \int_{-\infty}^{+\infty} f(ze^{-iu})e^{-|u|/t} du$ we have that $G_U^{\alpha,\beta}(f)(z)$ is analytic in $\mathbb{D}_R$ and we can write

$$G_U^{\alpha,\beta}(f)(z) = \sum_{k=0}^{\infty} a_k \cdot b_k(\alpha,\beta) \cdot z^k, z \in \mathbb{D}_R,$$

where $b_k(\alpha,\beta) = \frac{1}{Beta(\alpha,\beta)} \int_0^1 \frac{t^{\alpha-1}(1-t)^{\beta-1}}{1+t^2k^2}dt$.

Also, if f is not constant for $q = 0$ and not a polynomial of degree $\leq q-1$ for $q \in \mathbb{N}$, then for all $1 \leq r < r_1 < R$, $q \in \mathbb{N} \cup \{0\}$, $\alpha \in (0,\beta]$, we have

$$\|[G_U^{\alpha,\beta}(f)]^{(q)} - f^{(q)}\|_r \sim \alpha,$$

where the constants in the equivalence depend only on f, q, r, r_1, and β.

(iii) For $U_t(f)(z) = \frac{2t^3}{\pi} \int_{-\infty}^{+\infty} \frac{f(ze^{-iu})}{(u^2+t^2)^2} du$, we have that $G_U^{\alpha,\beta}(f)(z)$ is analytic in $\mathbb{D}_R$ and we can write

$$G_U^{\alpha,\beta}(f)(z) = \sum_{k=0}^{\infty} a_k \cdot b_k(\alpha,\beta) \cdot z^k, z \in \mathbb{D}_R,$$

where $b_k(\alpha,\beta) = \frac{1}{Beta(\alpha,\beta)} \int_0^1 t^{\alpha-1}(1-t)^{\beta-1}(1+kt)e^{-kt}dt$.

Also, if f is not constant for $q = 0$ and not a polynomial of degree $\leq q-1$ for $q \in \mathbb{N}$, then for all $1 \leq r < r_1 < R$, $q \in \mathbb{N} \cup \{0\}$, $\alpha \in (0,\beta]$, we have

$$\|[G_U^{\alpha,\beta}(f)]^{(q)} - f^{(q)}\|_r \sim \alpha,$$

where the constants in the equivalence depend only on f, q, r, r_1, and β.

(iv) For $U_t(f)(z) = \frac{1}{\sqrt{\pi t}} \int_{-\infty}^{+\infty} f(ze^{-iu})e^{-u^2/t} du$, we have that $G_U^{\alpha,\beta}(f)(z)$ is analytic in $\mathbb{D}_R$ and we can write

$$G_U^{\alpha,\beta}(f)(z) = \sum_{k=0}^{\infty} a_k \cdot b_k(\alpha,\beta) z^k, z \in \mathbb{D}_R,$$

where $b_k(\alpha,\beta) = \frac{1}{Beta(\alpha,\beta)} \int_0^1 t^{\alpha-1}(1-t)^{\beta-1}e^{-(k^2/4)t}dt$.

Also, if f is not constant for $q = 0$ and not a polynomial of degree $\leq q-1$ for $q \in \mathbb{N}$, then for all $1 \leq r < r_1 < R$, $q \in \mathbb{N} \cup \{0\}$, $\alpha \in (0,\beta]$, we have

$$\|[G_U^{\alpha,\beta}(f)]^{(q)} - f^{(q)}\|_r \sim \alpha,$$

where the constants in the equivalence depend only on f, q, r, r_1, and β.

Proof. (i) By Gal [49], p. 213, Theorem 3.2.5, (i), $U_t(f)(z)$ is analytic (as function of z) in $\mathbb{D}_R$ and we can write

$$U_t(f)(z) = \sum_{k=0}^{\infty} a_k e^{-kt} z^k, \text{ for all } |z| < R \text{ and } t \geq 0.$$

Since $|\sum_{k=0}^{\infty} a_k e^{-kt} z^k| \leq \sum_{k=0}^{\infty} |a_k| \cdot |z|^k < \infty$, this implies that for fixed $|z| < R$, the series in t, $\sum_{k=0}^{\infty} a_k e^{-kt} z^k$ is uniformly convergent on $[0, \infty)$, and therefore we immediately can write

$$G_U^{\alpha,\beta}(f)(z) = \sum_{k=0}^{\infty} a_k b_k(\alpha, \beta) z^k,$$

where

$$b_k(\alpha, \beta) = \frac{1}{Beta(\alpha, \beta)} \int_0^1 t^{\alpha-1}(1-t)^{\beta-1} e^{-kt} dt.$$

In other order of ideas, we easily can write

$$G_U^{\alpha,\beta}(f)(z) - f(z) = \frac{1}{Beta(\alpha, \beta)} \cdot \int_0^1 t^{\alpha-1}(1-t)^{\beta-1}[U_t(f)(z) - f(z)] dt,$$

which together with the estimate $|U_t(f)(z) - f(z)| \leq C_r(f)t$ in Gal [49], p. 213, Theorem 3.2.5, (iii), implies

$$|G_U^{\alpha,\beta}(f)(z) - f(z)| \leq \frac{1}{Beta(\alpha, \beta)} \cdot \int_0^1 t^{\alpha-1}(1-t)^{\beta-1}|U_t(f)(z) - f(z)| dt$$

$$\leq C_r(f)\frac{1}{Beta(\alpha, \beta)} \cdot \int_0^1 t^{\alpha}(1-t)^{\beta-1} dt = C_r(f) \cdot \frac{Beta(\alpha+1, \beta)}{Beta(\alpha, \beta)}$$

$$= C_r(f) \cdot \frac{\alpha}{\alpha+\beta} \leq C_r(f) \cdot \alpha,$$

for all $|z| \leq r$, where $C_r(f) > 0$ is independent of z (and α, β) but depends on f and r. Here we used the well-known formula $\frac{Beta(\alpha+1,\beta)}{Beta(\alpha,\beta)} = \frac{\alpha}{\alpha+\beta}$.

Now, for $q \in \mathbb{N} \cup \{0\}$ and $1 \leq r < r_1 < R$, by using Cauchy's formula and the standard reasonings in the proof of Theorem 2.2.1, we get the upper estimate

$$\|[G_U^{\alpha,\beta}(f)]^{(q)} - f^{(q)}\|_r \leq C^*\alpha,$$

with C^* depending only on f, q, r, and r_1.

It remains to prove the lower estimate. For this purpose, reasoning exactly as in the proof of Theorem 3.2.5, at pages 218–219 in the book of Gal [49], for $z = re^{i\varphi}$ and $p \in \mathbb{N} \cup \{0\}$, we get

$$\frac{1}{2\pi} \int_{-\pi}^{\pi} [f^{(q)}(z) - [U_t(f)]^{(q)}(z)] e^{-ip\varphi} d\varphi$$

$$= a_{q+p}(q+p)(q+p-1)\ldots(p+1)r^p[1 - e^{-(q+p)t}].$$

Multiplying above with $\frac{1}{Beta(\alpha,\beta)}t^{\alpha-1}(1-t)^{\beta-1}$ and then integrating with respect to t, it follows

$$I :=$$

$$\frac{1}{Beta(\alpha,\beta)} \cdot \int_0^1 \left\{ \frac{1}{2\pi} \int_{-\pi}^{\pi} [f^{(q)}(z) - [U_t(f)]^{(q)}(z)]e^{-ip\varphi}d\varphi \right\} t^{\alpha-1}(1-t)^{\beta-1}dt$$

$$= a_{q+p}(q+p)(q+p-1)\dots(p+1)r^p \frac{1}{Beta(\alpha,\beta)} \int_0^1 t^{\alpha-1}(1-t)^{\beta-1}[1-e^{-(q+p)t}]dt.$$

Applying Fubini's result to the double integral I and then passing to modulus, we easily obtain

$$\left| \frac{1}{2\pi} \int_{-\pi}^{\pi} e^{-ip\varphi} \left[\frac{1}{Beta(\alpha,\beta)} \int_0^1 [f^{(q)}(z) - [U_t(f)]^{(q)}(z)]t^{\alpha-1}(1-t)^{\beta-1}dt \right] d\varphi \right|$$

$$= |a_{q+p}|(q+p)(q+p-1)\dots(p+1)r^p$$

$$\cdot \left[\frac{1}{Beta(\alpha,\beta)} \int_0^1 t^{\alpha-1}(1-t)^{\beta-1}[1-e^{-(q+p)t}]dt \right].$$

Since

$$\frac{1}{Beta(\alpha,\beta)} \int_0^1 [f^{(q)}(z) - [U_t(f)]^{(q)}(z)]t^{\alpha-1}(1-t)^{\beta-1}dt$$

$$= f^{(q)}(z) - [G_U^{\alpha,\beta}(f)]^{(q)}(z),$$

the previous equality immediately implies

$$\left| \frac{1}{2\pi} \int_{-\pi}^{\pi} e^{-ip\varphi} \left[f^{(q)}(z) - (G_U^{\alpha,\beta}(f))^{(q)}(z) \right] d\varphi \right|$$

$$= |a_{q+p}|(q+p)(q+p-1)\dots(p+1)r^p$$

$$\cdot \left[\frac{1}{Beta(\alpha,\beta)} \int_0^1 t^{\alpha-1}(1-t)^{\beta-1}[1-e^{-(q+p)t}]dt \right]$$

and

$$|a_{q+p}|(q+p)(q+p-1)\dots(p+1)r^p$$

$$\cdot \left[\frac{1}{Beta(\alpha,\beta)} \int_0^1 t^{\alpha-1}(1-t)^{\beta-1}[1-e^{-(q+p)t}]dt \right]$$

$$\leq \|f^{(q)} - (G_U^{\alpha,\beta}(f))^{(q)}\|_r.$$

First take $q = 0$. In what follows, denoting

$$V_{\alpha,\beta} = \inf_{p\geq 1} \left(\frac{1}{Beta(\alpha,\beta)} \int_0^1 t^{\alpha-1}(1-t)^{\beta-1}[1-e^{-pt}]dt \right),$$

we clearly get

$$V_{\alpha,\beta} = \frac{1}{Beta(\alpha,\beta)} \int_0^1 t^{\alpha-1}(1-t)^{\beta-1}[1-e^{-t}]dt.$$

But denoting $g(t) = e^{-t}$, by the mean value theorem, there exists $\xi \in (0,1)$ such that $1 - e^{-t} = g(0) - g(t) = te^{-\xi} \geq \frac{t}{e}$, which immediately implies

$$V_{\alpha,\beta} \geq \frac{1}{e \cdot Beta(\alpha,\beta)} \int_0^1 t^{\alpha}(1-t)^{\beta-1}dt = \frac{Beta(\alpha+1,\beta)}{e \cdot Beta(\alpha,\beta)}$$

$$= \frac{1}{e} \cdot \frac{\alpha}{\alpha+\beta} \geq \frac{1}{e} \cdot \frac{\alpha}{2\beta} \geq \frac{\alpha}{2e}.$$

By following now for $q \geq 0$ the standard reasonings as in the proof of Theorem 2.2.1, we get the desired equivalence in the statement.

(ii) By Gal [49], p. 206, Theorem 3.2.1, (i), $U_t(f)(z)$ is analytic (as function of z) in $\mathbb{D}_R$ and we can write

$$U_t(f)(z) = \sum_{k=0}^{\infty} \frac{a_k}{1+t^2k^2}z^k, \text{ for all } |z| < R \text{ and } t \geq 0.$$

Since $|\sum_{k=0}^{\infty} \frac{a_k}{1+t^2k^2}z^k| \leq \sum_{k=0}^{\infty}|a_k| \cdot |z|^k < \infty$, this implies that for fixed $|z| < R$, the series in t, $\sum_{k=0}^{\infty} \frac{a_k}{1+t^2k^2}z^k$ is uniformly convergent on $[0,\infty)$, and therefore we immediately can write

$$G_U^{\alpha,\beta}(f)(z) = \sum_{k=0}^{\infty} a_k z^k \frac{1}{Beta(\alpha,\beta)} \int_0^1 \frac{t^{\alpha-1}(1-t)^{\beta-1}}{1+t^2k^2}dt.$$

In other order of ideas, we easily can write

$$G_U^{\alpha,\beta}(f)(z) - f(z) = \frac{1}{Beta(\alpha,\beta)} \cdot \int_0^1 t^{\alpha-1}(1-t)^{\beta-1}[U_t(f)(z) - f(z)]dt,$$

which together with the estimate $|U_t(f)(z) - f(z)| \leq C_r(f)t^2$ in Gal [49], p. 207, Theorem 3.2.1, (iv), implies

$$|G_U^{\alpha,\beta}(f)(z) - f(z)| \leq \frac{1}{Beta(\alpha,\beta)} \cdot \int_0^1 t^{\alpha-1}(1-t)^{\beta-1}|U_t(f)(z) - f(z)|dt$$

$$\leq C_r(f)\frac{1}{Beta(\alpha,\beta)} \cdot \int_0^1 t^{\alpha+1}(1-t)^{\beta-1}dt = C_r(f) \cdot \frac{Beta(\alpha+2,\beta)}{Beta(\alpha,\beta)}$$

$$= C_r(f)\frac{\alpha+1}{\alpha+\beta+1} \cdot \frac{\alpha}{\alpha+\beta} \leq C_r(f)\frac{\alpha(\alpha+1)}{2} \leq C_r(f)\alpha,$$

for all $|z| \leq r$, where $C_r(f) > 0$ is independent of z (and α, β) but depends on f and r.

Now, let $q \in \mathbb{N} \cup \{0\}$ and $1 \leq r < r_1 < R$. By standard reasonings and using Cauchy's formula as in the proof of Theorem 2.2.1, we get the upper estimate

$$\|[G_U^{\alpha,\beta}(f)]^{(q)} - f^{(q)}\|_r \leq C^*\alpha,$$

with C^* depending only on f, q, r, and r_1.

It remains to prove the lower estimate. For this purpose, reasoning exactly as in the proof of Theorem 3.2.1, at pages 209–210 in the book of Gal [49], for $z = re^{i\varphi}$ and $p \in \mathbb{N} \cup \{0\}$, we get

$$\frac{1}{2\pi} \int_{-\pi}^{\pi} [f^{(q)}(z) - [U_t(f)]^{(q)}(z)]e^{-ip\varphi}d\varphi$$

$$= a_{q+p}(q+p)(q+p-1)\dots(p+1)r^p \cdot \frac{t^2(q+p)^2}{1+t^2(q+p)^2}.$$

Multiplying above with $\frac{1}{Beta(\alpha,\beta)}t^{\alpha-1}(1-t)^{\beta-1}$ and then integrating with respect to t, it follows

$$I :=$$

$$\frac{1}{Beta(\alpha,\beta)} \cdot \int_0^1 \left\{ \frac{1}{2\pi} \int_{-\pi}^{\pi} [f^{(q)}(z) - [U_t(f)]^{(q)}(z)]e^{-ip\varphi}d\varphi \right\} t^{\alpha-1}(1-t)^{\beta-1}dt$$

$$= a_{q+p}(q+p)(q+p-1)\dots(p+1)r^p$$

$$\cdot \frac{1}{Beta(\alpha,\beta)} \int_0^1 t^{\alpha-1}(1-t)^{\beta-1}\left[\frac{t^2(q+p)^2}{1+t^2(q+p)^2} \right] dt.$$

Applying Fubini's result to the double integral I and then passing to modulus, we easily obtain

$$\left| \frac{1}{2\pi} \int_{-\pi}^{\pi} e^{-ip\varphi} \left[\frac{1}{Beta(\alpha,\beta)} \int_0^1 [f^{(q)}(z) - [U_t(f)]^{(q)}(z)]t^{\alpha-1}(1-t)^{\beta-1}dt \right] d\varphi \right|$$

$$= |a_{q+p}|(q+p)(q+p-1)\dots(p+1)r^p$$

$$\cdot \left[\frac{1}{Beta(\alpha,\beta)} \int_0^1 t^{\alpha-1}(1-t)^{\beta-1}\left[\frac{t^2(q+p)^2}{1+t^2(q+p)^2} \right] dt \right].$$

Since

$$\frac{1}{Beta(\alpha,\beta)} \int_0^1 [f^{(q)}(z) - [U_t(f)]^{(q)}(z)]t^{\alpha-1}(1-t)^{\beta-1}dt$$

$$= f^{(q)}(z) - [G_U^{\alpha,\beta}(f)]^{(q)}(z),$$

the previous equality immediately implies

$$\left| \frac{1}{2\pi} \int_{-\pi}^{\pi} e^{-ip\varphi} \left[f^{(q)}(z) - (G_U^{\alpha,\beta}(f))^{(q)}(z) \right] d\varphi \right|$$

$$= |a_{q+p}|(q+p)(q+p-1)\ldots(p+1)r^p$$

$$\cdot \left[\frac{1}{Beta(\alpha,\beta)} \int_0^1 t^{\alpha-1}(1-t)^{\beta-1} \left[\frac{t^2(q+p)^2}{1+t^2(q+p)^2} \right] dt \right]$$

and

$$|a_{q+p}|(q+p)(q+p-1)\ldots(p+1)r^p$$

$$\cdot \left[\frac{1}{Beta(\alpha,\beta)} \int_0^1 t^{\alpha-1}(1-t)^{\beta-1} \left[\frac{t^2(q+p)^2}{1+t^2(q+p)^2} \right] dt \right] \leq \|f^{(q)} - (G_U^{\alpha,\beta}(f))^{(q)}\|_r.$$

First take $q = 0$. From the previous inequality we immediately obtain

$$|a_p|r^p \left(\frac{1}{Beta(\alpha,\beta)} \int_0^1 t^{\alpha-1}(1-t)^{\beta-1} \left[\frac{t^2p^2}{1+t^2p^2} \right] dt \right)$$

$$\leq \|f - G_U^{\alpha,\beta}(f)\|_r.$$

In what follows, denoting

$$V_{\alpha,\beta} = \inf_{p\geq 1} \left(\frac{1}{Beta(\alpha,\beta)} \int_0^1 t^{\alpha-1}(1-t)^{\beta-1} \left[\frac{t^2p^2}{1+t^2p^2} \right] dt \right),$$

we clearly get

$$V_{\alpha,\beta} = \frac{1}{Beta(\alpha,\beta)} \int_0^1 t^{\alpha-1}(1-t)^{\beta-1} \left[\frac{t^2}{1+t^2} \right] dt$$

$$= \frac{1}{Beta(\alpha,\beta)} \int_0^1 t^{\alpha-1}(1-t)^{\beta-1} \left[1 - \frac{1}{1+t^2} \right] dt.$$

But we have $1 - \frac{1}{1+t^2} \geq \frac{t^2}{4}$, for all $t \in [0,1]$. Indeed, denoting $g(t) = 1 - \frac{1}{1+t^2} - \frac{t^2}{4}$, we get $g(0) = 0$ and $g'(t) = \frac{2t}{(1+t^2)^2} - \frac{2t}{4} = 2t\left(\frac{1}{(1+t^2)^2} - \frac{1}{4} \right) \geq 0$, for all $t \in [0,1]$. It follows that $g(t)$ is nondecreasing on $[0,1]$ and therefore $g(t) \geq 0$ for all $t \in [0,1]$.

In conclusion,

$$V_{\alpha,\beta} \geq \frac{1}{Beta(\alpha,\beta)} \int_0^1 t^{\alpha-1}(1-t)^{\beta-1}\frac{t^2}{4}dt$$

$$= \frac{1}{4} \cdot \frac{Beta(\alpha+2,\beta)}{Beta(\alpha,\beta)} = \frac{1}{4} \cdot \frac{\alpha+1}{\alpha+\beta+1} \cdot \frac{\alpha}{\alpha+\beta}$$

$$\geq \frac{1}{4} \cdot \frac{\alpha(\alpha+1)}{2} \geq \frac{\alpha}{8}.$$

By following now for $q \geq 0$ the standard reasonings as in the proof of Theorem 2.2.1, we get the desired equivalence in the statement.

(iii) By Gal [49], p. 213, Theorem 3.2.5, (i), $U_t(f)(z)$ is analytic (as function of z) in $\mathbb{D}_R$ and we can write

$$U_t(f)(z) = \sum_{k=0}^{\infty} a_k (1 + kt) e^{-kt} z^k, \text{ for all } |z| < R \text{ and } t \geq 0.$$

Since $|\sum_{k=0}^{\infty} a_k e^{-kt}(1+kt)z^k| \leq 2 \sum_{k=0}^{\infty} |a_k| \cdot |z|^k < \infty$, this implies that for fixed $|z| < R$, the series in t, $\sum_{k=0}^{\infty} a_k(1+kt)e^{-kt}z^k$ is uniformly convergent on $[0, \infty)$, and therefore we immediately can write

$$G_U^{\alpha,\beta}(f)(z) = \sum_{k=0}^{\infty} a_k z^k \frac{1}{Beta(\alpha, \beta)} \int_0^1 t^{\alpha-1}(1 - t)^{\beta-1}(1 + kt)e^{-kt} dt,$$

where denoting $b_k(\alpha, \beta) = \frac{1}{Beta(\alpha,\beta)} \int_0^1 t^{\alpha-1}(1-t)^{\beta-1}(1+kt)e^{-kt} dt$, we obtain

$$G_U^{\alpha,\alpha}(f)(z) = \sum_{k=0}^{\infty} a_k \cdot b_k(\alpha, \beta) \cdot z^k.$$

In other order of ideas, we easily can write

$$G_U^{\alpha,\beta}(f)(z) - f(z) = \frac{1}{Beta(\alpha, \beta)} \cdot \int_0^1 t^{\alpha-1}(1 - t)^{\beta-1}[U_t(f)(z) - f(z)] dt,$$

which together with the estimate $|U_t(f)(z) - f(z)| \leq C_r(f)t^2$ in Gal [49], p. 213–214, Theorem 3.2.5, (iv), implies

$$|G_U^{\alpha,\beta}(f)(z) - f(z)| \leq \frac{1}{Beta(\alpha, \beta)} \cdot \int_0^1 t^{\alpha-1}(1 - t)^{\beta-1}|U_t(f)(z) - f(z)| dt$$

$$\leq C_r(f)\frac{1}{Beta(\alpha, \beta)} \cdot \int_0^1 t^{\alpha+1}(1-t)^{\beta-1} dt = C_r(f) \cdot \frac{Beta(\alpha + 2, \beta)}{Beta(\alpha, \beta)} \leq C_r(f)\alpha,$$

for all $|z| \leq r$, where $C_r(f) > 0$ is independent of z (and α) but depends on f and r. We used here the estimate from the above point (ii).

Now, let $q \in \mathbb{N} \cup \{0\}$ and $1 \leq r < r_1 < R$. By standard reasonings and using Cauchy's formula as in the proof of Theorem 2.2.1, we get the upper estimate

$$\|[G_U^{\alpha,\beta}(f)]^{(q)} - f^{(q)}\|_r \leq C^*\alpha,$$

with C^* depending only on f, q, r, and r_1.

It remains to prove the lower estimate. For this purpose, reasoning exactly as in the proof of Theorem 3.2.5, at pages 219–220 in the book of Gal [49], for $z = re^{i\varphi}$ and $p \in \mathbb{N} \cup \{0\}$, we get

$$\frac{1}{2\pi} \int_{-\pi}^{\pi} [f^{(q)}(z) - [U_t(f)]^{(q)}(z)] e^{-ip\varphi} d\varphi$$

$$= a_{q+p}(q+p)(q+p-1)\ldots(p+1)r^p[1 - (1 + (q+p)t)e^{-(q+p)t}].$$

Multiplying above with $\frac{1}{Beta(\alpha,\beta)} t^{\alpha-1}(1-t)^{\beta-1}$ and then integrating with respect to t, it follows

$$I :=$$

$$\frac{1}{Beta(\alpha,\beta)} \cdot \int_0^1 \left\{ \frac{1}{2\pi} \int_{-\pi}^{\pi} [f^{(q)}(z) - [U_t(f)]^{(q)}(z)] e^{-ip\varphi} d\varphi \right\} t^{\alpha-1}(1-t)^{\beta-1} dt$$

$$= a_{q+p}(q+p)(q+p-1)\ldots(p+1)r^p$$

$$\cdot \frac{1}{Beta(\alpha,\beta)} \int_0^1 t^{\alpha-1}(1-t)^{\beta-1} \left[1 - (1 + (q+p)t)e^{-(q+p)t} \right] dt.$$

Applying Fubini's result to the double integral I and then passing to modulus, we easily obtain

$$\left| \frac{1}{2\pi} \int_{-\pi}^{\pi} e^{-ip\varphi} \left[\frac{1}{Beta(\alpha,\beta)} \int_0^1 [f^{(q)}(z) - [U_t(f)]^{(q)}(z)] t^{\alpha-1}(1-t)^{\beta-1} dt \right] d\varphi \right|$$

$$= |a_{q+p}|(q+p)(q+p-1)\ldots(p+1)r^p$$

$$\cdot \left[\frac{1}{Beta(\alpha,\beta)} \int_0^1 t^{\alpha-1}(1-t)^{\beta-1} \left[1 - (1 + (q+p)t)e^{-(q+p)t} \right] dt \right].$$

Since

$$\frac{1}{Beta(\alpha,\beta)} \int_0^1 [f^{(q)}(z) - [U_t(f)]^{(q)}(z)] t^{\alpha-1}(1-t)^{\beta-1} dt$$

$$= f^{(q)}(z) - [G_U^{\alpha,\beta}(f)]^{(q)}(z),$$

the previous equality immediately implies

$$\left| \frac{1}{2\pi} \int_{-\pi}^{\pi} e^{-ip\varphi} \left[f^{(q)}(z) - (G_U^{\alpha,\beta}(f))^{(q)}(z) \right] d\varphi \right|$$

$$= |a_{q+p}|(q+p)(q+p-1)\ldots(p+1)r^p$$

$$\cdot \left[\frac{1}{Beta(\alpha,\beta)} \int_0^1 t^{\alpha-1}(1-t)^{\beta-1} \left[1 - (1 + (q+p)t)e^{-(q+p)t} \right] dt \right]$$

and

$$|a_{q+p}|(q+p)(q+p-1)\ldots(p+1)r^p$$

$$\cdot \left[\frac{1}{Beta(\alpha,\beta)} \int_0^1 t^{\alpha-1}(1-t)^{\beta-1} \left[1 - (1 + (q+p)t)e^{-(q+p)t} \right] dt \right]$$

$$\leq \| f^{(q)} - (G_U^{\alpha,\beta}(f))^{(q)} \|_r.$$

First take $q = 0$. From the previous inequality, we immediately obtain

$$|a_p| r^p \left(\frac{1}{Beta(\alpha, \beta)} \int_0^1 t^{\alpha-1}(1-t)^{\beta-1} \left[1 - (1 + pt)e^{-pt} \right] dt \right)$$

$$\leq \| f - G_U^{\alpha,\beta}(f) \|_r.$$

In what follows, denoting

$$V_{\alpha,\beta} = \inf_{p \geq 1} \left(\frac{1}{Beta(\alpha, \beta)} \int_0^1 t^{\alpha-1}(1-t)^{\beta-1} \left[1 - (1 + pt)e^{-pt} \right] dt \right),$$

we immediately get

$$V_{\alpha,\beta} = \frac{1}{Beta(\alpha, \beta)} \int_0^1 t^{\alpha-1}(1-t)^{\beta-1} \left[1 - (1 + t)e^{-t} \right] dt.$$

But we have $1 - (1 + t)e^{-t} \geq \frac{t^2}{e}$, for all $t \in [0, 1]$. Indeed, denoting $g(t) = 1 - (1+t)e^{-t} - \frac{t^2}{e}$, we have $g(0) = 0$ and $g'(t) = te^{-t} - \frac{t}{e} = t\left(\frac{1}{e^t} - \frac{1}{e} \right) \geq 0$ for all $t \in [0, 1]$. This implies that $g(t)$ is nondecreasing on $[0, 1]$ and therefore $g(t) \geq 0$ for all $t \in [0, 1]$.

Therefore,

$$V_{\alpha,\beta} \geq \frac{1}{Beta(\alpha, \beta)} \int_0^1 t^{\alpha-1}(1-t)^{\beta-1}\frac{t^2}{2e} dt$$

$$= \frac{Beta(\alpha + 2, \beta)}{2e \cdot B(\alpha, \beta)} = \frac{1}{2e} \cdot \frac{\alpha+1}{\alpha+\beta+1} \cdot \frac{\alpha}{\alpha+\beta}$$

$$\geq \frac{1}{2e} \cdot \frac{\alpha(\alpha+1)}{2} \geq \frac{\alpha}{4e}.$$

By following now for $q \geq 0$ the standard reasonings as in the proof of Theorem 2.4.1, we get the desired equivalence in the statement.

(iv) By Gal [49], p. 223, Theorem 3.2.8, (i), $U_t(f)(z)$ is analytic (as function of z) in $\mathbb{D}_R$ and we can write

$$U_t(f)(z) = \sum_{k=0}^{\infty} a_k e^{-k^2 t/4} z^k, \quad \text{for all } |z| < R \text{ and } t \geq 0.$$

Since $\left| \sum_{k=0}^{\infty} a_k e^{-k^2 t/4} z^k \right| \leq \sum_{k=0}^{\infty} |a_k| \cdot |z|^k < \infty$, this implies that for fixed $|z| < R$, the series in t, $\sum_{k=0}^{\infty} a_k e^{-k^2 t/4} z^k$ is uniformly convergent on $[0, \infty)$, and therefore we immediately can write

$$G_U^{\alpha,\beta}(f)(z) = \sum_{k=0}^{\infty} a_k z^k \frac{1}{Beta(\alpha, \beta)} \int_0^1 t^{\alpha-1}(1-t)^{\beta-1}e^{-(k^2/4)t} dt,$$

where denoting $b_k(\alpha, \beta) = \frac{1}{Beta(\alpha,\beta)} \int_0^1 t^{\alpha-1}(1-t)^{\beta-1}e^{-(k^2/4)t}dt$, we can write

$$G_U^{\alpha,\beta}(f)(z) = \sum_{k=0}^{\infty} a_k \cdot b_k(\alpha, \beta) \cdot z^k.$$

In other order of ideas, we easily can write

$$G_U^{\alpha,\beta}(f)(z) - f(z) = \frac{1}{Beta(\alpha, \beta)} \cdot \int_0^1 t^{\alpha-1}(1 - t)^{\beta-1}[U_t(f)(z) - f(z)]dt,$$

which together with the estimate $|U_t(f)(z) - f(z)| \leq C_r(f)t$ in Gal [49], p. 224, Theorem 3.2.8, (iv), implies

$$|G_U^{\alpha,\beta}(f)(z) - f(z)| \leq \frac{1}{Beta(\alpha, \beta)} \cdot \int_0^1 t^{\alpha-1}(1 - t)^{\beta-1}|U_t(f)(z) - f(z)|dt$$

$$\leq C_r(f)\frac{1}{Beta(\alpha, \beta)} \cdot \int_0^1 t^{\alpha}(1 - t)^{\beta-1}dt = C_r(f) \cdot \frac{Beta(\alpha + 1, \beta)}{Beta(\alpha, \beta)} \leq C_r(f)\alpha,$$

for all $|z| \leq r$, where $C_r(f) > 0$ is independent of z (and α) but depends on f and r.

Now, let $q \in \mathbb{N} \cup \{0\}$ and $1 \leq r < r_1 < R$. By standard reasonings and using Cauchy's formula as in the proof of Theorem 2.2.1, we get the upper estimate

$$\|[G_U^{\alpha,\beta}(f)]^{(q)} - f^{(q)}\|_r \leq C^*\alpha,$$

with C^* depending only on f, q, r, and r_1.

It remains to prove the lower estimate. For this purpose, reasoning exactly as in the proof of Theorem 3.2.8, at pages 227–228 in the book of Gal [49], for $z = re^{i\varphi}$ and $p \in \mathbb{N} \cup \{0\}$, we get

$$\frac{1}{2\pi} \int_{-\pi}^{\pi} [f^{(q)}(z) - [U_t(f)]^{(q)}(z)]e^{-ip\varphi}d\varphi$$

$$= a_{q+p}(q + p)(q + p - 1) \dots (p + 1)r^p[1 - e^{-(q+p)^2t/4}].$$

Multiplying above with $\frac{1}{Beta(\alpha,\beta)}t^{\alpha-1}(1 - t)^{\beta-1}$ and then integrating with respect to t, it follows

$$I :=$$

$$\frac{1}{Beta(\alpha, \beta)} \cdot \int_0^1 \left\{ \frac{1}{2\pi} \int_{-\pi}^{\pi} [f^{(q)}(z) - [U_t(f)]^{(q)}(z)]e^{-ip\varphi}d\varphi \right\} t^{\alpha-1})1 - t)^{\beta-1}dt$$

$$= a_{q+p}(q + p)(q + p - 1) \dots (p + 1)r^p$$

$$\cdot \frac{1}{Beta(\alpha, \beta)} \int_0^1 t^{\alpha-1}(1 - t)^{\beta-1} \left[1 - e^{-(q+p)^2t/4} \right] dt.$$

Applying Fubini's result to the double integral I and then passing to modulus, we easily obtain

$$\left| \frac{1}{2\pi} \int_{-\pi}^{\pi} e^{-ip\varphi} \left[\frac{1}{Beta(\alpha,\beta)} \int_{0}^{\infty} [f^{(q)}(z) - [U_t(f)]^{(q)}(z)] t^{\alpha-1}(1-t)^{\beta-1} dt \right] d\varphi \right|$$

$$= |a_{q+p}|(q+p)(q+p-1)\ldots(p+1)r^p$$

$$\cdot \left[\frac{1}{Beta(\alpha,\beta)} \int_{0}^{1} t^{\alpha-1}e^{-t} \left[1 - e^{-(q+p)^2 t/4} \right] dt \right].$$

Since

$$\frac{1}{Beta(\alpha,\beta)} \int_{0}^{1} [f^{(q)}(z) - [U_t(f)]^{(q)}(z)] t^{\alpha-1}(1-t)^{\beta-1} dt$$

$$= f^{(q)}(z) - [G_U^{\alpha,\beta}(f)]^{(q)}(z),$$

the previous equality immediately implies

$$\left| \frac{1}{2\pi} \int_{-\pi}^{\pi} e^{-ip\varphi} \left[f^{(q)}(z) - (G_U^{\alpha,\beta}(f))^{(q)}(z) \right] d\varphi \right|$$

$$= |a_{q+p}|(q+p)(q+p-1)\ldots(p+1)r^p$$

$$\cdot \left[\frac{1}{Beta(\alpha,\beta)} \int_{0}^{1} t^{\alpha-1}(1-t)^{\beta-1} \left[1 - e^{-(q+p)^2 t/4} \right] dt \right]$$

and

$$|a_{q+p}|(q+p)(q+p-1)\ldots(p+1)r^p$$

$$\cdot \left[\frac{1}{Beta(\alpha,\beta)} \int_{0}^{1} t^{\alpha-1}(1-t)^{\beta-1} \left[1 - e^{-(q+p)^2 t/4} \right] dt \right]$$

$$\leq \|f^{(q)} - (G_U^{\alpha,\beta}(f))^{(q)}\|_r.$$

First take $q = 0$. From the previous inequality, we immediately obtain

$$|a_p|r^p \left(\frac{1}{Beta(\alpha,\beta)} \int_{0}^{1} t^{\alpha-1}(1-t)^{\beta-1} \left[1 - e^{-p^2 t/4} \right] dt \right) \leq \|f - G_U^{\alpha,\beta}(f)\|_r.$$

In what follows, denoting

$$V_{\alpha,\beta} = \inf_{p \geq 1} \left(\frac{1}{Beta(\alpha,\beta)} \int_{0}^{1} t^{\alpha-1}(1-t)^{\beta-1} \left[1 - e^{-p^2 t/4} \right] dt \right),$$

by simple calculation, we get

$$V_{\alpha,\beta} = \frac{1}{Beta(\alpha,\beta)} \int_{0}^{1} t^{\alpha-1}(1-t)^{\beta-1} \left[1 - e^{-t/4} \right] dt.$$

But denoting $g(t) = e^{-t/4}$, by the mean value theorem, there exists $\xi \in (0,1)$ such that $1 - e^{-t/4} = g(0) - g(t) = t\frac{e^{-\xi/4}}{4} \geq \frac{t}{4e^{1/4}}$, which immediately implies

$$V_{\alpha,\beta} \geq \frac{1}{4e^{1/4} \cdot Beta(\alpha,\beta)} \int_0^1 t^\alpha (1-t)^{\beta-1} dt = \frac{Beta(\alpha+1,\beta)}{4e^{1/4} \cdot Beta(\alpha,\beta)}$$

$$= \frac{1}{4e^{1/4}} \cdot \frac{\alpha}{\alpha+\beta} \geq \frac{1}{4e^{1/4}} \cdot \frac{\alpha}{2\beta} \geq \frac{\alpha}{8e^{1/4}}.$$

By following now for $q \geq 0$ the standard reasonings as in the proof of Theorem 2.2.1, we get the desired equivalence in the statement.

The theorem is proved. □

Concerning the overconvergence phenomenon for the potentials of real variable x, $F_U^\alpha(f)(x)$ and $G_U^{\alpha,\beta}(f)(x)$, we can present the next two results.

Theorem 2.2.3. *Let $d > 0$ and suppose that $f : S_d \to \mathbb{C}$ is bounded and uniformly continuous in the strip $S_d = \{z = x + iy \in \mathbb{C}; x \in \mathbb{R}, |y| \leq d\}$.*

(i) *Denoting $U_t(f)(z) = \frac{1}{2t} \int_{-\infty}^{+\infty} f(z+u)e^{-|u|/t} du$, for all $0 < \alpha \leq 1/2$ and $z \in S_d$, we have*

$$|F_U^\alpha(f)(z) - f(z)| \leq 5\frac{\alpha+1}{\alpha}\omega_2(f;\alpha)_{S_d},$$

where

$$\omega_2(f;\delta)_{S_d} = \sup\{|f(u+t) - 2f(u) + f(u-t)|; u, u-t, u+t \in S_d, |t| \leq \delta\}.$$

(ii) *Denoting $U_t(f)(z) = \frac{2t^3}{\pi} \int_{-\infty}^{+\infty} \frac{f(z+u)}{(u^2+t^2)^2} du$, for all $0 < \alpha \leq 1/2$ and $z \in S_d$, we have*

$$|F_U^\alpha(f)(z) - f(z)| \leq C\frac{\alpha+1}{\alpha}\omega_2(f;\alpha)_{S_d},$$

where $C > 0$ is independent of z, α, and f.

(iii) *Denoting $U_t(f)(z) = \frac{1}{\sqrt{\pi t^2}} \int_{-\infty}^{+\infty} f(z+u)e^{-u^2/t^2} dt$, for all $0 < \alpha \leq 1/2$ and $z \in S_d$, we have*

$$|F_U^\alpha(f)(z) - f(z)| \leq C\frac{\alpha+1}{\alpha}\omega_2(f;\alpha)_{S_d},$$

where $C > 0$ is independent of z, α, and f.

Proof. (i) If $z \in S_d$ then clearly that for all $t \in \mathbb{R}$, we have $z + t \in S_d$, and since f is bounded in S_d (denote its bound by $M(f)$), it easily follows $|U_t(f)(z)| \leq 2M(f)$ for all $z \in S_d$. Therefore $U_t(f)(z)$ exists for all $z \in S_d$. Also, the uniform continuity of f on S_d implies that $0 \leq \lim_{t \to 0} \omega_2(f;t)_{S_d} \leq 2\lim_{t \to 0} \omega_1(f;t)_{S_d} = 0$.

For all $z \in S_d$, we have

$$|U_t(f)(z) - f(z)| = \left| \frac{1}{2t} \int_0^\infty [f(z+u) - 2f(z) + f(z-u)]e^{-|u|/t}du \right|$$

$$\leq \frac{1}{2t} \int_0^\infty \omega_2(f;(u/t)t)_{S_d} e^{-u/t}du$$

$$\leq \omega_2(f;t)_{S_d} \frac{1}{2t} \int_0^\infty [1+(u/t)]^2 e^{-u/t}du = \frac{5}{2}\omega_2(f;t)_{S_d}.$$

For the last equality, see Gal [48], pp. 252–253, proof of Theorem 5.2. We get

$$|F_U^\alpha(f)(z) - f(z)| \leq \frac{1}{\Gamma(\alpha)} \int_0^\infty t^{\alpha-1}e^{-t}|U_t(f)(z) - f(z)|dt$$

$$\leq \frac{5}{2} \cdot \frac{1}{\Gamma(\alpha)} \int_0^\infty t^{\alpha-1}e^{-t}\omega_2(f;t)_{S_d}dt = \frac{5}{2} \cdot \frac{1}{\Gamma(\alpha)} \int_0^\infty t^{\alpha-1}e^{-t}\omega_2(f;\alpha(t/\alpha))_{S_d}dt$$

$$\leq \frac{5}{2} \cdot \omega_2(f;\alpha)_{S_d} \cdot \frac{1}{\Gamma(\alpha)} \int_0^\infty t^{\alpha-1}e^{-t}[1+(t/\alpha)]^2dt$$

$$= \frac{5}{2} \cdot \omega_2(f;\alpha)_{S_d} \left[1 + \frac{2}{\Gamma(\alpha)\alpha} \int_0^\infty t^\alpha e^{-t}dt + \frac{1}{\Gamma(\alpha)\alpha^2} \int_0^\infty t^{\alpha+1}e^{-t}dt \right]$$

$$= \frac{5}{2} \cdot \omega_2(f;\alpha)_{S_d} \left[1 + \frac{2\Gamma(\alpha+1)}{\Gamma(\alpha)\alpha} + \frac{\Gamma(\alpha+2)}{\Gamma(\alpha)\alpha^2} \right] = \frac{5}{2} \cdot \omega_2(f;\alpha)_{S_d} \left[3 + \frac{\alpha+1}{\alpha} \right]$$

$$\leq 5\frac{\alpha+1}{\alpha}\omega_2(f;\alpha)_{S_d},$$

because $3 \leq \frac{\alpha+1}{\alpha}$.

(ii) We obtain

$$|U_t(f)(z) - f(z)| =$$

$$\left| \frac{2t^3}{\pi} \int_0^\infty \frac{[f(z+u) - 2f(z) + f(z-u)]}{(u^2+t^2)^2}du \right| \leq \frac{2t^3}{\pi} \int_0^{+\infty} \frac{\omega_2(f;(u/t)t)_{S_d}}{(u^2+t^2)^2}du$$

$$\leq \omega_2(f;t)_{S_d} \frac{2t^3}{\pi} \int_0^\infty \left[1 + \frac{u}{t}\right]^2 \cdot \frac{1}{(u^2+t^2)^2}du \leq C\omega_2(f;t)_{S_d},$$

since by easy calculation, we get that

$$\frac{2t^3}{\pi} \int_0^\infty \left[1 + \frac{u}{t}\right]^2 \cdot \frac{1}{(u^2+t^2)^2}du \leq C,$$

where $C > 0$ is independent of t, z, and f.

Reasoning exactly as at the above point (i), we obtain the estimate

$$|F_U^\alpha(f)(z) - f(z)| \leq C \frac{\alpha+1}{\alpha} \omega_2(f;\alpha)_{S_d},$$

where $C > 0$ is independent of f, z, and α.

(iii) We get

$$|U_t(f)(z) - f(z)| = \left| \frac{1}{\sqrt{\pi t}} \int_0^\infty [f(z+u) - 2f(z) + f(z-u)] e^{-u^2/t} du \right| \leq$$

$$\frac{1}{\sqrt{\pi t}} \int_0^{+\infty} \omega_2(f; (u/\sqrt{t})\sqrt{t})_{S_d} e^{-u^2/t} du$$

$$\leq \omega_2(f;\sqrt{t})_{S_d} \frac{1}{\sqrt{\pi t}} \int_0^\infty \left[\frac{u}{\sqrt{t}} + 1 \right]^2 e^{-u^2/t} du \leq C\omega_2(f;\sqrt{t})_{S_d},$$

since

$$\frac{1}{\sqrt{\pi t}} \int_0^\infty \frac{u^2}{t} e^{-u^2/t} du = \frac{1}{\sqrt{\pi}} \int_0^\infty v^2 e^{-v^2} dv < \infty$$

and

$$\frac{2}{\sqrt{\pi t}} \int_0^\infty \frac{u}{\sqrt{t}} e^{-u^2/t} du = \frac{2}{\sqrt{\pi t}} \sqrt{t} \int_0^\infty v e^{-v^2} dv = \frac{2}{\sqrt{\pi}} \int_0^\infty v e^{-v^2} dv < \infty.$$

Above, $C > 0$ is independent of t, z, and f.

Reasoning exactly as at the above point (i), we obtain the estimate

$$|F_U^\alpha(f)(z) - f(z)| \leq C \frac{\alpha+1}{\alpha} \omega_2(f;\alpha)_{S_d},$$

where $C > 0$ is independent of f, z, and α.

The theorem is proved. $\square$

Remark. If f is such that its second derivative f'' is bounded in the strip S_d, then by the mean value theorem in complex analysis (see, e.g., Stancu [133], p. 258, Exercise 4.20), we get $\omega_2(f;\alpha)_{S_d} \leq C\alpha^2$, and therefore it follows the upper estimate $\frac{\alpha+1}{\alpha}\omega_2(f;\alpha)_{S_d} \leq C_1\alpha$, which proves the overconvergence phenomenon as $\alpha \to 0$ for $F^\alpha(f)(x)$ in all the three cases for $U_t(f)(z)$.

Theorem 2.2.4. *Let $d > 0$ and suppose that $f : S_d \to \mathbb{C}$ is bounded and uniformly continuous in the strip $S_d = \{z = x + iy \in \mathbb{C}; x \in \mathbb{R}, |y| \leq d\}$. Also, suppose that $0 < \alpha \leq \beta \leq 1$, $\alpha + \beta \geq 1$.*

(i) Denoting $U_t(f)(z) = \frac{1}{2t} \int_{-\infty}^{+\infty} f(z+u) e^{-|u|/t} du$, for all $0 < \alpha < 1$ and $z \in S_d$, we have

$$|G_U^{\alpha,\beta}(f)(z) - f(z)| \leq 5 \frac{\alpha+1}{\alpha} \omega_2(f;\alpha)_{S_d},$$

where

$$\omega_2(f;\delta)_{S_d} = \sup\{|f(u+t) - 2f(u) + f(u-t)|; u, u-t, u+t \in S_d, |t| \leq \delta\}.$$

(ii) Denoting $U_t(f)(z) = \frac{2t^3}{\pi} \int_{-\infty}^{+\infty} \frac{f(z+u)}{(u^2+t^2)^2} du$, for all $0 < \alpha < 1$ and $z \in S_d$, we have

$$|G_U^{\alpha,\beta}(f)(z) - f(z)| \leq C\frac{\alpha+1}{\alpha}\omega_2(f;\alpha)_{S_d},$$

where $C > 0$ is independent of z, α, and f.
(iii) Denoting $U_t(f)(z) = \frac{1}{\sqrt{\pi t^2}} \int_{-\infty}^{+\infty} f(z+u)e^{-u^2/t^2} du$, for all $0 < \alpha < 1$ and $z \in S_d$, we have

$$|G_U^{\alpha,\beta}(f)(z) - f(z)| \leq C\frac{\alpha+1}{\alpha} \cdot \omega_2(f;\sqrt{\alpha})_{S_d},$$

where $C > 0$ is independent of z, α, and f.

Proof. (i) If $z \in S_d$ then clearly that for all $t \in \mathbb{R}$, we have $z + t \in S_d$, and since f is bounded in S_d (denote its bound by $M(f)$), it easily follows $|U_t(f)(z)| \leq 2M(f)$ for all $z \in S_d$. Therefore $U_t(f)(z)$ exists for all $z \in S_d$. Also, the uniform continuity of f on S_d implies that $0 \leq \lim_{\xi \to 0} \omega_2(f;t)_{S_d} \leq 2\lim_{t \to 0} \omega_1(f;t)_{S_d} = 0$.

As in the proof of the above Theorem 2.2.3, (i), we have

$$|U_t(f)(z) - f(z)| \leq \frac{5}{2} \cdot \omega_2(f;t)_{S_d}, \text{ for all } z \in S_d.$$

We get

$$|G_U^{\alpha,\beta}(f)(z) - f(z)| \leq \frac{1}{Beta(\alpha,\beta)} \int_0^1 t^{\alpha-1}(1-t)^{\beta-1}|U_t(f)(z) - f(z)|dt$$

$$\leq \frac{5}{2} \cdot \frac{1}{Beta(\alpha,\beta)} \int_0^1 t^{\alpha-1}(1-t)^{\beta-1}\omega_2(f;t)_{S_d}dt$$

$$= \frac{5}{2} \cdot \frac{1}{Beta(\alpha,\beta)} \int_0^1 t^{\alpha-1}(1-t)^{\beta-1}\omega_2(f;\alpha(t/\alpha))_{S_d}dt$$

$$\leq \frac{5}{2} \cdot \omega_2(f;\alpha)_{S_d} \cdot \frac{1}{Beta(\alpha,\beta)} \int_0^1 t^{\alpha-1}(1-t)^{\beta-1}[1+(t/\alpha)]^2dt$$

$$= \frac{5}{2} \cdot \omega_2(f;\alpha)_{S_d}$$

$$\left[1 + \frac{2}{Beta(\alpha,\beta)\alpha} \int_0^1 t^\alpha(1-t)^{\beta-1}dt + \frac{1}{Beta(\alpha,\beta)\alpha^2} \int_0^1 t^{\alpha+1}(1-t)^{\beta-1}dt\right]$$

$$= \frac{5}{2} \cdot \omega_2(f;\alpha)_{S_d} \left[1 + \frac{2Beta(\alpha+1,\beta)}{Beta(\alpha,\beta)\alpha} + \frac{Beta(\alpha+2,\beta)}{Beta(\alpha,\beta)\alpha^2}\right]$$

$$= \frac{5}{2} \cdot \omega_2(f;\alpha)_{S_d} \left[1 + \frac{2}{\alpha} \cdot \frac{\alpha}{\alpha+\beta} + \frac{1}{\alpha^2} \cdot \frac{\alpha+1}{\alpha+\beta+1} \cdot \frac{\alpha}{\alpha+\beta} \right]$$

$$\leq \frac{5}{2} \cdot \omega_2(f;\alpha)_{S_d} \left[3 + \frac{\alpha+1}{2\alpha} \right] \leq 5\frac{\alpha+1}{\alpha}\omega_2(f;\alpha)_{S_d},$$

because $1 \leq \frac{\alpha+1}{2\alpha}$.

(ii) As in the proof of the above Theorem 2.2.3, (ii), we have

$$|U_t(f)(z) - f(z)| \leq C\omega_2(f;t)_{S_d}, \text{ for all } z \in S_d, t \geq 0,$$

where $C > 0$ is independent of t, z, and f.

Reasoning exactly as at the above point (i), we obtain the estimate

$$|G_U^{\alpha,\beta}(f)(z) - f(z)| \leq C\frac{\alpha+1}{\alpha}\omega_2(f;\alpha)_{S_d},$$

where $C > 0$ is independent of f, z, and α.

(iii) As in the proof of the above Theorem 2.2.3, (iii), we have

$$|U_t(f)(z) - f(z)| \leq C\omega_2(f;t)_{S_d}, \text{ for all } z \in S_d, t \geq 0,$$

where $C > 0$ is independent of t, z, and f.

Reasoning exactly as at the above point (i), we obtain the estimate

$$|G_U^{\alpha,\beta}(f)(z) - f(z)| \leq C\frac{\alpha+1}{\alpha}\omega_2(f;\alpha)_{S_d},$$

where $C > 0$ is independent of f, z, and α.

The theorem is proved.   □

Remarks. 1) If f is such that its second derivative f'' is bounded in the strip S_d, then by the mean value theorem in complex analysis (see, e.g., Stancu [133], p. 258, Exercise 4.20), we get $\omega_2(f;\alpha)_{S_d} \leq C\alpha^2$, and therefore it follows the upper estimate $\frac{\alpha+1}{\alpha}\omega_2(f;\alpha)_{S_d} \leq C_1\alpha$, which proves the overconvergence phenomenon as $\alpha \to 0$ for $G^{\alpha,\beta}(f)(x)$ in all the three cases for $U_t(f)(z)$.

2) Note that Theorem 2.2.1 differs from Theorem 2.2.3 and Theorem 2.2.2 differs from Theorem 2.2.4, by the different formulas for the corresponding $U_t(f)(z)$. Thus, in Theorems 2.2.1 and 2.2.2, these $U_t(f)(z)$ are complex convolution-type integrals, while in Theorems 2.2.3 and 2.2.4, they are obtained from their real correspondents, simply replacing the real variable x by the complex one z.

At the end of this section, we study the approximation properties of the complex Bessel-type potential

$$B^\alpha(f)(z,t) = \frac{1}{\Gamma(\alpha/2)} \int_0^\infty \left[\int_{-\infty}^\infty \tau^{(\alpha/2)-1} e^{-\tau} W(y,\tau^2) f(ze^{-iy}, t-\tau) dy \right] d\tau,$$

where $\Gamma(\alpha)$ is the Gamma function and

$$W(y,\tau^2) = \frac{1}{\sqrt{4\pi\tau^2}} e^{-y^2/(4\tau^2)} = \frac{1}{2\tau\sqrt{\pi}} e^{-y^2/(4\tau^2)}.$$

In this sense, we present the following:

Theorem 2.2.5. *Let us suppose that the function $f(z,t)$ is in $L^p(\overline{\mathbb{D}_1} \times \mathbb{R})$, $1 \le p < \infty$, where $\mathbb{D}_1 = \{z \in \mathbb{C}; |z| < 1\}$ and that $f(z,t)$ is analytic in $\mathbb{D}_1$ and continuous in $\overline{\mathbb{D}_1}$ for any fixed $t \in \mathbb{R}$. Then for all $|z| \le 1$, $t \in \mathbb{R}$ and $\alpha > 0$, we have*

$$|B^\alpha(f)(z,t) - f(z,t)| \le \left(2 + \frac{2}{\sqrt{\pi}} \right) \omega_1(f; \alpha/2, \alpha/2),$$

where

$$\omega_1(f; a, b)$$
$$= \sup\{|f(u,t) - f(v,s)|; |u-v| \le a; u,v \in \overline{\mathbb{D}_1}, |t-s| \le b; t,s \in \mathbb{R}\}.$$

Proof. We have

$$B^\alpha(f)(z,t) - f(z,t)$$

$$= \frac{1}{\Gamma(\alpha/2)} \int_0^\infty \left[\int_{-\infty}^\infty \tau^{(\alpha/2)-1} e^{-\tau} W(y,\tau^2)(f(ze^{-iy}, t-\tau) - f(z,t)) dy \right] d\tau$$

$$= \frac{1}{\Gamma(\alpha/2)} \int_0^\infty$$

$$\cdot \left[\int_0^\infty \tau^{(\alpha/2)-1} e^{-\tau} W(y,\tau^2)(f(ze^{-iy}, t-\tau) - f(z,t) \right.$$

$$\left. + f(ze^{iy}, t-\tau) - f(z,t)) dy \right] d\tau.$$

In all what follows, for the simplicity of notations, denote $a = \alpha/2$. Passing to absolute value for $|z| \le 1$, we immediately get

$$|B^\alpha(f)(z,t) - f(z,t)|$$

$$\le \frac{1}{\Gamma(a)} \int_0^\infty \left[2 \cdot \int_0^\infty \tau^{a-1} e^{-\tau} W(y,\tau^2) \omega_1(f; y, \tau) dy \right] d\tau$$

$$= \frac{1}{\Gamma(a)} \int_0^\infty \left[2 \cdot \int_0^\infty \tau^{a-1} e^{-\tau} W(y,\tau^2) \omega_1(f; a(y/a), a(\tau/a)) dy \right] d\tau$$

$$= \omega_1(f; a, a) \cdot \frac{1}{\Gamma(a)} \int_0^\infty \tau^{a-1} e^{-\tau} \left[2 \cdot \int_0^\infty \left[1 + \frac{y}{a} + \frac{\tau}{a} \right] W(y, \tau^2) dy \right] d\tau$$

$$= \omega_1(f; a, a) \cdot \frac{1}{\Gamma(a)} \int_0^\infty \tau^{a-1} e^{-\tau} \left[1 + \frac{\tau}{a} + \frac{2\tau}{\sqrt{\pi}a} \right] d\tau = \left(2 + \frac{2}{\sqrt{\pi}} \right) \omega_1(f; a, a).$$

We used here the simple formulas $\int_0^\infty y W(y, \tau^2) dy = \frac{\tau}{\sqrt{\pi}}$ and $\Gamma(a+1) = a\Gamma(a)$.

The theorem is proved. $\square$

2.3 Notes

Note 2.3.1. Theorems 2.1.1, 2.2.3, 2.2.4, and 2.2.5 appear for the first time here.

Chapter 3

Overconvergence in ℂ of the Orthogonal Expansions

This chapter mainly studies the overconvergence phenomenon in compact sets in ℂ, of the orthogonal expansions attached to the interval $[-1, 1]$.

The overconvergence phenomenon in ℂ for the orthogonal expansions attached to complex-valued functions defined on real intervals is well known. For example, in the cases of Chebyshev and Legendre orthogonal expansions on $[-1, 1]$, attached to an analytic function in an ellipse of the complex plane with foci at 1 an -1, it is well known that in any closed subset of that ellipse, the partial sums converge uniformly to f, with a geometric rate of convergence, but without explicit constants (see, e.g., Davis [30], pp. 89–90, Lemma 4.4.2 there and Theorem 12.4.7). The main goal of this chapter is to present a systematic study of this aspect, by obtaining explicit constants and explicit geometric orders of approximation. It is worth noting that the results are obtained in the more general setting of Banach space-valued functions.

The study begins in Sect. 3.1 with the problem of convergence for the Chebyshev orthogonal expansion attached to a vector-valued function defined on $[-1, 1]$. In the next two sections, we treat the overconvergence phenomenon in ℂ for the orthogonal expansions attached to Banach space-valued functions. Thus, Sect. 3.2 contains the general results for any kind of orthogonal expansions, while in Sect. 3.3 we apply these results to the orthogonal expansions of Chebyshev and Legendre kinds. Sect. 3.4 presents with details some interesting open problems concerning the possibility of applying the results obtained to other orthogonal polynomials too, such as to the Hermite polynomials, Laguerre polynomials, or Gegenbauer polynomials.

S.G. Gal, *Overconvergence in Complex Approximation*,
DOI 10.1007/978-1-4614-7098-4_3, © Springer Science+Business Media New York 2013

3.1 Convergence of the Chebyshev Orthogonal Expansions for Vector-Valued Functions

By using a nice and powerful method based on a classical result in functional analysis, in the recent book of Gal [49], pp. 299–314, basic results in the approximation of vector-valued functions of real variable by polynomials with coefficients in normed spaces, called generalized polynomials, were obtained. More exactly, in the case of vector-valued functions of a real variable, estimates in terms of Ditzian–Totik L^p-moduli of smoothness and inverse theorems for approximation by Bernstein, Bernstein–Kantorovich, and Szász–Mirakjan, Baskakov generalized operators and their Kantorovich analogues, and Post–Widder and Jackson-type generalized operators, were obtained. Also, in the same book of Gal [49], pp. 314–321, in the case of vector-valued functions of a complex variable, quantitative estimates for Bernstein, Butzer, de la Vallée–Poussin, Riesz–Zygmund, Jackson, Poisson–Cauchy, Gauss–Weierstrass, q-Picard, and q-Gauss–Weierstrass operators were proved. Note here that because we can take as vector space the space of all complex numbers $\mathbb{C}$, it follows that in the case of real variable too, one frames into the title of the present book. In this section we continue this kind of study, by obtaining the quantitative approximation of vector-valued functions of real variable and of complex variable, by Chebyshev orthogonal expansions.

For our purpose we need the following ingredients.

Let $(X, ||\cdot||)$ be a normed space over K, where $K = \mathbb{R}$ or $K = \mathbb{C}$. It is well known that absolutely identical to the case of real- (or complex-) valued functions, the following concepts in the Definitions 3.1.1–3.1.3 can be introduced.

Definition 3.1.1. (i) Let $(X, ||\cdot||)$ be a normed space over K, where $K = \mathbb{R}$ or $K = \mathbb{C}$.

A generalized algebraic polynomial of degree $\leq n$, with coefficients in X, will be an expression of the form $P_n(x) = \sum_{k=0}^{n} c_k x^k$, where $c_k \in X, k = 0, \ldots, n$ and $x \in [a, b]$.

(ii) Denote by $\mathcal{P}_n[a, b]$, $\mathcal{T}_n$, $\mathcal{P}_n(K)$ the sets of all generalized algebraic, trigonometric, and complex polynomials of degree $\leq n$ with coefficients in X, respectively. Also, for $1 \leq p \leq +\infty$ and $f : [a, b] \to X$ or $f : R \to X$ or $f : D \to X$ ($D = \{z \in \mathbb{C}; |z| < 1\}$), we define the quantity of best approximation by

$$E_n(f)_p = \inf\{||f - P||_p; P \in \mathcal{P}_n[a, b]\},$$

where $||f||_\infty = \sup_x\{||f(x)||\}$ and $||f||_p = (\int_a^b ||f(x)||^p dx)^{1/p}$, if $f : [a, b] \to X$.

Also, if $||f||_\infty < \infty$, then we write that $f \in C([a, b]; X)$, and if $||f||_p < \infty$, $1 \leq p < \infty$, we write $f \in L^p([a, b]; X)$.

Definition 3.1.2. $f : [a, b] \to X$ will be called Riemann integrable on $[a, b]$, if there exists an element $I \in X$ denoted by $\int_a^b f(x)dx$, with the following property: for any $\varepsilon > 0$, there exists $\delta > 0$, such that for any division of $[a, b]$, $d : a = x_0 < \ldots < x_n = b$ with the norm $\nu(d) < \delta$ and any intermediary points $\xi_i \in [x_i, x_{i+1}]$, we have $||S(f; d, \xi_i) - I|| < \varepsilon$, where $S(f; d, \xi_i) = \sum_{i=0}^{n-1} f(\xi_i)(x_{i+1} - x_i)$.

Denote by $L^p([a, b]; X) = \{f : [a, b] \to X; f$ is pth Bochner–Lebesgue integrable and $\int_a^b ||f(x)||^p dx < +\infty\}, 1 \le p < \infty$, where the equality between two functions in $L^p([a, b]; X)$ is considered in the almost everywhere sense. For $p = +\infty$, we consider $L^p([a, b]; X) = C([a, b]; X)$.

Definition 3.1.3. For $f : [a, b] \to X$, the kth L^p-modulus of smoothness of f on $[a, b]$ will be given by

$$\omega_k(f; \delta)_p = \sup\left\{ \left(\int_a^{b-kh} ||\Delta_h^k f(x)||^p dx \right)^{1/p}; 0 \le h \le \delta \right\}, \text{ if } 1 \le p < +\infty,$$

and

$$\omega_k(f; \delta)_\infty = \sup_{0 \le h \le \delta} \{\sup\{||\Delta_h^k f(x)||; x, x + kh \in [a, b]\}\}.$$

Here $\Delta_h^k f(x) = \sum_{j=0}^n (-1)^{k-j} \binom{k}{j} f(x + jh)$.

The main tool used in our proof is based on the following well-known result in functional analysis.

Theorem 3.1.4 (see, e.g., Muntean [115], p. 183). *Let $(X, || \cdot ||)$ be a normed space over K, the real or the complex numbers, and denote by X^* the conjugate space of X. Then, $||x|| = \sup\{|x^*(x)| : x^* \in X^*, |||x^*||| \le 1\}$, for all $x \in X$.*

Let $\{T_0/\sqrt{2}, T_1, \ldots, T_n, \ldots\}$ be the system of Chebyshev polynomials, i.e., $T_n(x) = \cos[n \arccos(x)]$, which is orthonormal with respect to the scalar product $< f, g > = \frac{2}{\pi} \int_{-1}^1 f(x)g(x) \frac{1}{\sqrt{1-x^2}} dx$.

For $f : [-1, 1] \to X$, continuous on $[-1, 1]$, where $(X, || \cdot ||)$ is a normed space, let us consider the coefficients

$$A_n = \frac{2}{\pi} \int_{-1}^1 f(x)T_n(x) \frac{1}{\sqrt{1-x^2}} dx = \frac{2}{\pi} \int_0^\pi f(\cos(\theta))\cos(k\theta)d\theta, n \ge 1,$$

$$A_0 = \frac{1}{\pi} \int_0^\pi f(\cos(\theta))d\theta.$$

Obviously $A_n \in X, \forall n \ge 0$, and for f, we can attach its Chebyshev expansions $S_n(f)(x) = \sum_{k=0}^n A_k T_k(x) \in X$, for all $n \in \mathbb{N}$.

The main result is the following.

Theorem 3.1.5 (Gal [54]).

(i) If the Dini–Lipschitz condition

$$\lim_{\delta \to 0} \omega_1(f; \delta) \log \delta = 0$$

holds, then

$$\lim_{n \to \infty} ||S_n(f) - f||_\infty = 0,$$

where $||f||_\infty = \sup\{||f(x)||; x \in [-1,1]\}$.
(ii) $E_n(f)_\infty \leq ||A_{n+1}|| + ||A_{n+2}|| + \dots$.

Proof. (i) For $x^* : X \to K$ linear and continuous with $|||x^*||| \leq 1$ and for $f : [-1,1] \to X$, continuous on X, let us define $g : [-1,1] \to K$ by $g(x) = x^*[f(x)]$, which obviously is continuous on $[-1,1]$.

We have $|g(x) - g(y)| = |x^*[f(x)] - x^*[f(y)]| \leq |||x^*||| \cdot ||f(x) - f(y)|| \leq ||f(x) - f(y)||$, wherefrom passing to supremum with $|x - y| \leq \delta$, we get $\omega_1(g; \delta) \leq \omega_1(f; \delta)$, for all $\delta \geq 0$.

Then $\omega_1(g; \delta) \log \delta \leq \omega_1(f; \delta) \log \delta$, which by hypothesis immediately implies $\lim_{\delta \to 0} \omega_1(g; \delta) \log \delta = 0$.

From, e.g., Cheney [25], p. 129, it follows that $S_n(g)(x) = \sum_{k=0}^{n} a_k T_k(x)$ converges uniformly to $g(x)$ on $[-1,1]$, where $a_k = \frac{2}{\pi} \int_{-1}^{1} g(x) T_k(x) \frac{1}{\sqrt{1-x^2}} dx$.

This is a consequence of the relation (see, e.g., Cheney [25], p. 147)

$$||S_n(g) - g||_\infty \leq C log(n)\omega_1(g; \frac{1}{n}),$$

where $C > 0$ is an absolute constant.

Taking into account the relation between $\omega_1(g; \delta)$ and $\omega_1(f; \delta)$ and that from the continuity and linearity of x^*, we have $S_n(g)(x) - g(x) = x^*[S_n(f)(x) - f(x)]$, we obtain

$$|x^*[S_n(f)(x) - f(x)]| \leq C log(n)\omega_1(f; \frac{1}{n}),$$

for all $x \in [1,1]$ and all x^* with $|||x^*||| \leq 1$.

Passing to supremum with x^* and taking into account Theorem 3.1.4, we immediately get

$$||S_n(f)(x) - f(x)|| \leq C log(n)\omega_1(f; \frac{1}{n}), \forall x \in [-1,1],$$

which by hypothesis implies

$$\lim_{n \to \infty} ||S_n(f) - f||_\infty = 0.$$

(ii) Keeping the notations from the above point (i) and taking into account, e.g., Cheney [25], p. 131, proof of Theorem 4 and statement of Theorem 5, we have

$$||S_n(g) - g||_\infty \leq |a_{n+1}| + |a_{n+2}| + \ldots + .$$

From the obvious relation $a_k = x^*(A_k)$, we get $|a_k| \leq |||x^*||| \cdot ||A_k|| \leq ||A_k||$ and therefore

$$|x^*[S_n(f)(x) - f(x)]| \leq ||A_{n+1}|| + ||A_{n+1}|| + \ldots +,$$

for all $x \in [-1, 1]$ and x^* with $|||x^*||| \leq 1$. Passing now to supremum with $|||x^*||| \leq 1$ and taking into account Theorem 3.1.4 too, we obtain

$$||S_n(f)(x) - f(x)|| \leq ||A_{n+1}|| + ||A_{n+1}|| + \ldots +,$$

for all $x \in [-1, 1]$.

Now, since $E_n(f)_\infty \leq ||S_n(f) - f||_\infty$, we immediately get the desired conclusion. $\qquad\square$

3.2 Overconvergence in $\mathbb{C}$ of the Orthogonal Expansions for Vector-Valued Functions

In this section we obtain very explicit quantitative estimates for the overconvergence in the complex plane of the partial sums of the Fourier-type expansions on $[-1, 1]$ with respect to orthogonal systems, for functions defined on $[-1, 1]$ and with values in a complex Banach space.

First we recall some well-known concepts and results for real-valued functions.

Let $I \subset \mathbb{R}$ be an open real interval (bounded or unbounded) and $\rho : I \to \mathbb{R}_+$ be a weight function, that is, continuous and positive on I such that all the integrals $\int_I x^k \rho(x)dx$, $k = 0, 1, 2, \ldots$ exist (finite). If we define $L_\rho^2(I) = \{f : I \to \mathbb{R}; \int_I \rho(x)|f(x)|^2 dx < \infty\}$, then $< f, g >_\rho = \int_I \rho(x)f(x)g(x)dx$ becomes an inner product and $(L_\rho^2(I), < \cdot, \cdot >_\rho)$ is a Hilbert space.

A sequence of polynomials $(Q_n)_{n \in \mathbb{N} \cup \{0\}}$ with degree $(Q_n) = n$ is called orthogonal on I with respect to the weight ρ if $< Q_n, Q_m >_\rho = 0$ if and only if $n \neq m$. Denoting $||f||_\rho = \sqrt{< f, f >_\rho}$ and $P_n(x) = \frac{Q_n(x)}{||Q_n||_\rho}$, the sequence $(P_n)_{n \in \mathbb{N} \cup \{0\}}$ becomes an orthonormal sequence on I.

Given $f \in L_\rho^2(I)$, $a_k(f) = < f, P_k >_\rho$, $k = 0, 1, 2, \ldots$, are called the coefficients of f with respect to $(P_n)_n$ and the series $\sum_{k=0}^\infty a_k(f)P_n(x)$ is called the expansion of f on I with respect to $(P_n)_n$.

For $x \in I$, qualitative and quantitative approximation results of the above expansion and of its partial sums $\sum_{k=0}^n a_k(f)P_n(x), n \in \mathbb{N}$, are well known.

On the other hand, it is a natural question to ask what overconvergence properties have the complex partial sums $\sum_{k=0}^n a_k(f)P_n(z)$ for $z \in \mathbb{C}$, that is, simply replacing $x \in I \subset \mathbb{R}$ by $z \in G \subset \mathbb{C}$, where $I \subset G$ and f is supposed analytic in G?

Note that $a_k(f), k = 0, 1, 2, \ldots$, remain here calculated on the real interval I and that the above expansion $\sum_{k=0}^{n} a_k(f) P_n(z)$ with $z \in \mathbb{C}$ evidently is different from the expansion in the complex plane with respect to the same system $(P_k(z))_k$, but with respect to another inner product in the form of a complex integral $[f, g]_\rho = \int_E f(z)\overline{g(z)}|\rho(z)| \cdot |dz|$.

In the case of the overconvergence phenomenon for Chebyshev and Legendre expansions on $[-1, 1]$, qualitative results and quantitative estimates but without explicit constants (in an ellipse of analyticity for f) are well known (see, e.g., Boyd [23], Theorem 2, p. 35, and Theorem 10, p. 52, and Davis [30], Theorem 12.4.7 and Lemma 4.4.2, together with their proofs).

The aim of the present section is to obtain very explicit quantitative estimates (concerning both the order and the constants) in the overconvergence of the orthogonal expansions on $[-1, 1]$ with respect to general orthogonal polynomials. The results will be obtained in the more general setting of vector-valued functions. In this sense, we will need the following additional considerations.

Definition 3.2.1 (see, e.g., Hille–Phillips [83], p. 92–93). *Let $(X, \|\cdot\|)$ be a complex Banach space, $R > 1$, and $f : \overline{\mathbb{D}}_R \to X$. We say that f is analytic (holomorphic) on $\mathbb{D}_R$ if for any $x^* \in B_1 = \{x^* : X \to \mathbb{C}; x^*$ linear and continuous, $\||x^*\|| \leq 1\}$, the function $g : \overline{\mathbb{D}}_R \to \mathbb{C}$ given by $g(z) = x^*[f(z)]$ is analytic on $\mathbb{D}_R$. (Here $\||\cdot\||$ represents the usual norm in the dual space X^*.)*

We denote by $A(\overline{\mathbb{D}}_R; X)$ the space of all functions $f : \overline{\mathbb{D}}_R \to X$ which are continuous on $\overline{\mathbb{D}}_R$ and analytic (holomorphic) on $\mathbb{D}_R$. It is a Banach space with respect to the norm $\|f\|_R = \max\{\|f(z)\|; z \in \overline{\mathbb{D}}_R\}$.

Theorem 3.2.2 (see, e.g., Hille–Phillips [83], p. 93). *Let $(X, \|\cdot\|)$ be a complex Banach space. If $f : \mathbb{D}_R \to X$ is analytic (holomorphic) on $\mathbb{D}_R$, then $f(z)$ is continuous (as mapping between two metric spaces) and differentiable in the sense that $f'(z) \in \mathbb{C}$ exists given by*

$$\lim_{h \to 0} \left\| \frac{f(z+h) - f(z)}{h} - f'(z) \right\| = 0,$$

uniformly with respect to z in any compact subset of $\mathbb{D}_R$.

Theorem 3.2.3 (see, e.g., Hille–Phillips [83], p. 97). *Let $(X, \|\cdot\|)$ be a complex Banach space. If $f : \mathbb{D}_R \to X$ is analytic (holomorphic) on $\mathbb{D}_R$, then we have the Taylor expansion*

$$f(z) = \sum_{n=0}^{\infty} \frac{f^{(n)}(0)}{n!} z^n, \quad z \in \mathbb{D}_R,$$

where the series converges uniformly on any compact subset of $\mathbb{D}_R$.

For $f : I \to X$, $\rho : I \to \mathbb{R}_+$ where $I \subset \mathbb{R}$ is an open real interval, $(X, \|\cdot\|)$ is a normed space and for $(P_n)_{n \in \mathbb{N} \cup \{0\}}$ a sequence of orthonormal polynomials on I with respect to the weight function $\rho : I \to \mathbb{R}_+$, we can attach the coefficients

$$a_k(f) = <f, P_k>_{\rho,X} = \int_I \rho(x) f(x) P_k(x) dx \in X, \; k = 0, 1, 2, \dots,$$

the partial sums $\sum_{k=0}^{n} a_k(f) P_k(x) \in X$, $n \in \mathbb{N} \cup \{0\}$, and the expansion of f on I with respect to the orthonormal sequence $(P_k)_k$ given by $\sum_{k=0}^{\infty} a_k(f) P_k(x)$.

When $X = \mathbb{R}$ or $X = \mathbb{C}$, evidently that we recapture the considerations from the beginning of this section.

In what follows, we provide two general results on the overconvergence of orthogonal expansions of a vector-valued function which could be used for any particular system of orthogonal functions.

The first main result is the following.

Theorem 3.2.4. Let $(X, \|\cdot\|)$ be a complex Banach space.

(i) For a bounded interval $I \subset \mathbb{R}$ and $\mathbb{D}_R = \{z \in \mathbb{C}; |z| < R\}$ such that $I \subset \mathbb{D}_R$, let us consider a sequence of polynomials $(P_n)_{n \in \mathbb{N} \cup \{0\}}$ orthonormal on I with respect to the weight ρ. If $f : \mathbb{D}_R \to X$ is analytic in $\mathbb{D}_R$, that is, (see Theorem 3.2.3) $f(z) = \sum_{j=0}^{\infty} c_j z^j$ for all $z \in \mathbb{D}_R$, where $c_k \in X$ for all k, then denoting $S_n(f)(z) = \sum_{k=0}^{n} <f, P_k>_{\rho,X} \cdot P_k(z)$ and $e_j(z) = z^j$, we have $S_n(f)(z) = \sum_{j=0}^{\infty} c_j S_n(e_j)(z)$ for all $z \in \mathbb{D}_R$.

(ii) Suppose that $I \subset \mathbb{R}$ is $I = (0, \infty)$ or $I = (-\infty, +\infty)$ and that $G = \mathbb{D}_R \bigcup I$. Also, let us consider a sequence of polynomials $(P_n)_{n \in \mathbb{N} \cup \{0\}}$ orthonormal on I with respect to the weight ρ.

If $f : G \to X$ is analytic on $\mathbb{D}_R$ (i.e., $f(z) = \sum_{j=0}^{\infty} c_j z^j$ for all $z \in \mathbb{D}_R$) and $f \in L_\rho^2(I; X)$ (see Definition 3.1.2), then

$$S_n(f)(z) = \sum_{k=0}^{n} <f, P_k>_\rho \cdot P_k(z) = \sum_{j=0}^{\infty} c_j S_n(e_j)(z), \; \text{for all } z \in \mathbb{D}_R.$$

Proof. (i) Denote $f_m(z) = \sum_{j=0}^{m} c_j e_j(z)$, $m \in \mathbb{N}$ and take $0 < r < R$ such that $I \subset \mathbb{D}_r$. By the hypothesis, it follows that $S_n(f_m)(z)$ is well defined for all $n, m \in \mathbb{N}$.

Let $x^* \in X^*$ be with $\|\|x^*\|\| \leq 1$, arbitrary. Note that if we define $g(z) = x^*[f(z)]$ and $g_m(z) = x^*[f_m(z)]$, then evidently that $g : \mathbb{D}_R \to \mathbb{C}$ is analytic in $\mathbb{D}_R$ and the linearity and continuity of x^* immediately implies $g_m(z) = \sum_{j=0}^{m} x^*(c_j) e_j(z)$ and $g(z) = \sum_{j=0}^{m} x^*(c_j) e_j(z)$ for all $z \in \mathbb{D}_R$.

Taking into account that by the concept of integral in Definition 3.1.2 we easily get $x^* \left[\int_I \rho(x) f(x) P_k(x) dx \right] = \int_I \rho(x) x^*[f(x)] P_k(x) dx$, it follows

$$|x^*[S_n(f)(z) - S_n(f_m)(z)]| = |x^*[S_n(f)(z)] - x^*[S_n(f_m)(z)]|$$

$$= \left| \sum_{k=0}^{n} x^*[< f - f_m, P_k >_{\rho,X}] \cdot P_k(z) \right| = \left| \sum_{k=0}^{n} < x^*[f - f_m], P_k >_{\rho} \cdot P_k(z) \right|$$

which by the Cauchy–Schwarz inequality $| < F, G >_{\rho} | \leq \|F\|_{\rho} \cdot \|G\|_{\rho}$ implies

$$\leq \sum_{k=0}^{n} \|x^*(f - f_m)\|_{\rho} \cdot \|P_k\|_{\rho} \cdot |P_k(z)|.$$

But since

$$\|x^*(f) - x^*(f_m)\|_{\rho} \leq \|x^*(f) - x^*(f_m)\|_{C(\overline{I})} \cdot \left(\int_{I} \rho(x)dx \right)^{1/2}$$

$$\leq c_{\rho} \|x^*(f) - x^*(f_m)\|_{C(\overline{\mathbb{D}}_r)} \leq c_{\rho} \||x^*\|| \cdot \|f - f_m\|_{C(\overline{\mathbb{D}}_r, X)} \leq c_{\rho} \cdot \|f - f_m\|_{C(\overline{\mathbb{D}}_r, X)},$$

for all $z \in \overline{\mathbb{D}}_r$, it follows

$$|x^*[S_n(f)(z) - S_n(f_m)(z)]| \leq c_{\rho} \|f - f_m\|_{C(\overline{\mathbb{D}}_r, X)} \sum_{k=0}^{n} |P_k(z)|$$

$$\leq c_{\rho,n,r} \|f - f_m\|_{C(\overline{\mathbb{D}}_r, X)}.$$

Here $\| \cdot \|_{C(\overline{\mathbb{D}}_r, X)}$ denotes the uniform norm in the standard space

$$C(\overline{\mathbb{D}}_r, X) = \{f : \overline{\mathbb{D}}_r \to X : f \text{ is continuous on } \overline{\mathbb{D}}_r\}.$$

Passing to supremum with x^* and taking into account Theorem 3.2.4, we obtain

$$\|S_n(f)(z) - S_n(f_m)(z)\| \leq c_{\rho,n,r} \|f - f_m\|_{C(\overline{\mathbb{D}}_r, X)}, \text{ for all } z \in \overline{\mathbb{D}}_r.$$

Since $\|f - f_m\|_{C(\overline{\mathbb{D}}_r, X)} \to 0$ as $m \to \infty$, passing above to limit with $m \to \infty$ (n being fixed), we obtain

$$\lim_{m \to \infty} S_n(f_m)(z) = \lim_{m \to \infty} \left\{ \sum_{k=0}^{n} < \sum_{j=0}^{m} c_j e_j, P_k >_{\rho,X} \cdot P_k(z) \right\} =$$

$$\lim_{m \to \infty} \left\{ \sum_{j=0}^{m} \left[\sum_{k=0}^{n} < c_j e_j, P_k >_{\rho,X} \cdot P_k(z) \right] \right\} = \lim_{m \to \infty} \sum_{j=0}^{m} c_j S_n(e_j)(z) =$$

$$\sum_{j=0}^{\infty} c_j S_n(e_j)(z) = S_n(f)(z),$$

which proves the point (i).

(ii) We will indicate the proof only in the case of $I = (0, \infty)$, since the case $I = (-\infty, +\infty)$ is similar.

Let $0 < r < R$. For any $m \in \mathbb{N}$, define

$$f_m(z) = \sum_{j=0}^{m} c_j z^j \text{ if } |z| \le r \text{ and } f_m(x) = f(x) \text{ if } x \in (r, +\infty).$$

(In the case when $I = (-\infty, +\infty)$, we define

$$f_m(z) = \sum_{j=0}^{m} c_j z^j \text{ if } |z| \le r \text{ and } f_m(x) = f(x) \text{ if } x \in (-\infty, -r) \bigcup (r, +\infty).)$$

By the hypothesis, it follows that $S_n(f_m)(z)$ is well defined for all $n, m \in \mathbb{N}$.

Let $x^* \in X^*$ be with $|||x^*||| \le 1$, arbitrary. Reasoning as in the proof of the above point (i), we get

$$|x^*[S_n(f)(z) - S_n(f_m)(z)]| = \left| \sum_{k=0}^{n} < x^*(f - f_m), P_k >_{\rho,X} \cdot P_k(z) \right| \le$$

$$\sum_{k=0}^{n} \|x^*(f - f_m)\|_\rho \cdot \|P_k\|_\rho \cdot |P_k(z)|.$$

But since

$$\|x^*(f - f_m)\|_\rho = \left(\int_I \rho(x) |x^*(f(x) - f_m(x))|^2 dx \right)^{1/2} =$$

$$\left(\int_0^r \rho(x) |x^*(f(x) - f_m(x))|^2 dx \right)^{1/2} \le \|x^*(f-f_m)\|_{C([0,r])} \cdot \left(\int_I \rho(x) dx \right)^{1/2} \le$$

$$c_\rho \|x^*(f - f_m)\|_{\overline{\mathbb{D}}_r} \le c_\rho |||x^*||| \cdot \|f - f_m\|_{C(\overline{\mathbb{D}}_r;X)} \le c_\rho \cdot \|f - f_m\|_{C(\overline{\mathbb{D}}_r;X)},$$

continuing the reasonings as in the proof of the above point (i), we finally arrive at the desired conclusion. □

An important consequence of Theorem 3.2.4 is the second main result of this section.

Corollary 3.2.5. *Suppose the hypothesis in Theorem 3.2.4, (i), or in Theorem 3.2.4, (ii), holds true. If $e_j(z) = \sum_{q=0}^{j} C_{j,q} P_q(z)$, for all $j = 0, 1, 2, \ldots$, (where $C_{j,q} \in \mathbb{R}$) then for all $z \in \mathbb{D}_R$, we have*

$$\|S_n(f)(z) - f(z)\| \le \sum_{j=n+1}^{\infty} \|c_j\| \cdot \sum_{q=n+1}^{j} |C_{j,q}| \cdot |P_q(z)|.$$

Proof. Firstly, note that by Theorem 3.2.4, (i) and (ii), we can write

$$\|S_n(f)(z) - f(z)\| \leq \sum_{j=0}^{\infty} \|c_j\| \cdot |S_n(e_j)(z) - e_j(z)|,$$

where $S_n(e_j)(z)$ denotes the partial sum of order n of the orthogonal expansion of the function $e_j(z) = z^k$, with respect to the orthonormal sequence of polynomials $(P_k)_k$.

In what follows, we prove that

$$S_n(e_j)(z) = e_j(z), \text{ for all } 0 \leq j \leq n \text{ and } z \in \mathbb{C}.$$

Indeed, we get

$$S_n(e_j)(z) = \sum_{k=0}^{n} <e_j, P_k>_\rho \cdot P_k(z) = \sum_{k=0}^{n}\sum_{q=0}^{j} C_{j,q} \cdot <P_q, P_k>_\rho \cdot P_k(z) =$$

$$\sum_{q=0}^{j} C_{j,q} P_q(z) = e_j(z).$$

Now suppose that $j > n$. We obtain

$$S_n(e_j)(z) = \sum_{k=0}^{n}\sum_{q=0}^{j} C_{j,q} \cdot <P_q, P_k>_\rho \cdot P_k(z) = \sum_{q=0}^{n} C_{j,q} P_q(z),$$

which implies $S_n(e_j)(z) - e_j(z) = \sum_{q=n+1}^{j} C_{j,q} P_q(z)$.

By Theorem 3.2.4, (i) or (ii), it follows

$$\|S_n(f)(z) - f(z)\| \leq \sum_{j=0}^{n} \|c_j\| \cdot |S_n(e_j)(z) - e_j(z)| + \sum_{j=n+1}^{\infty} \|c_j\| \cdot |S_n(f)(e_j)(z) - e_j(z)|$$

$$= \sum_{j=n+1}^{\infty} \|c_j\| \cdot |S_n(f)(e_j)(z) - e_j(z)| \leq \sum_{j=n+1}^{\infty} \|c_j\| \cdot \sum_{q=n+1}^{j} |C_{j,q}| \cdot |P_q(z)|,$$

which proves the theorem. $\square$

3.3 Applications to Chebyshev and Legendre Orthogonal Expansions

In what follows we apply the Corollary 3.2.5 to the particular cases of Chebyshev and Legendre orthogonal polynomials. In the first application one refers to the orthogonal Chebyshev polynomials given by $T_k(x) = \cos[k \arccos x]$, $k = 0, 1, 2, \ldots,$. First we briefly recall the history on the overconvergence of Chebyshev series. The associated orthonormal system becomes $P_k(x) = \frac{\sqrt{2}}{\sqrt{\pi}} T_k(x)$ for $k \geq 1$ and $P_0(x) = \frac{1}{\sqrt{\pi}} T_0(x) = \frac{1}{\sqrt{\pi}}$. In this case, it is well known that $I = (-1,1)$ and the weight is given by $\rho(x) = \frac{1}{\sqrt{1-x^2}}$. Concerning the convergence properties, according to, e.g., Theorem 5.16, p. 130 in Mason and Handscomb [107], if the complex-valued function f, of complex variable, is analytic in the ellipse of semiaxis $a = \frac{r+r^{-1}}{2}$ and $b = \frac{r-r^{-1}}{2}$ (where $r > 1$) with foci at -1 and 1, then

$$|f(x) - S_n(f)(x)| \le \frac{M}{r^n}, \text{ for all } x \in [-1,1].$$

Also, by Theorem 7, p. 48 in Boyd [23], the Chebyshev series

$$\sum_{k=0}^{\infty} a_k(f) P_k(z)$$

converges to $f(z)$ in the interior of an ellipse where f is supposed to be analytic.

Regarding now the order of convergence, by, e.g., Davis [30], p. 90, Exercise 9, if in addition we would have $\limsup_{n\to\infty} |a_n|^{1/n} = \frac{1}{r}$, with $1 < r \le \infty$, then by taking into account the Lemma 4.4.2 in Davis [30], p. 89–90, it would follow the overconvergence (in the interior of an ellipse where f is supposed to be analytic) with the order of convergence of a geometric series, but still without a completely explicit estimate for the order and for the constants (see the proof of Lemma 4.4.2 in Davis [30]).

Note that since by Cheney [25], p. 157, proof of Theorem, for arbitrary f, we have that

$$|a_n(f)| \le \frac{\|f''\|_{C[-1,1]}}{n^2}, \text{ for all } n \in \mathbb{N},$$

this immediately implies that $\limsup_{n\to\infty} |a_n|^{1/n} = 1 = \frac{1}{r}$, so that the above-mentioned reasonings cannot be applied.

However, by using Corollary 3.2.5, we will obtain a rate of convergence in the ellipse E_r with $r > 1$, with very explicit estimates, for the more general case of normed valued functions, as follows.

Theorem 3.3.1. *Let $(X, \|\cdot\|)$ be a complex Banach space. Suppose that $f : \mathbb{D}_R \to X$ is analytic in $\mathbb{D}_R$ with $R > 1$, that is, $f(z) = \sum_{j=0}^{\infty} c_j z^j$ for all $z \in \mathbb{D}_R$, with $c_j \in X$, and defined by*

$$S_n(f)(z) = \sum_{k=0}^{n} <f, P_k>_{\rho,X} \cdot P_k(z),$$

the partial sum of the Chebyshev series attached to f, where $P_k(x), k = 0, 1, 2, \ldots$, denotes the sequence of orthonormal Chebyshev polynomials defined at the beginning of this section. Also, denote by E_r the ellipse with the semiaxis $a = \frac{r+r^{-1}}{2}$ and $b = \frac{r-r^{-1}}{2}$ (where $r > 1$), with foci at -1 and 1.

(i) For any r satisfying $1 < r < R$, the rate of convergence is given by

$$\|S_n(f)(z) - f(z)\| \le \frac{2r\sqrt{2}}{r-1} \sum_{j=n+1}^{\infty} \|c_j\| r^j,$$

for all $z \in int\,(E_r) \bigcup E_r$ and $n \in \mathbb{N}$, where $\sum_{j=0}^{\infty} \|c_j\| r^j < \infty$.

(ii) Let $m \in \mathbb{N}$, $m \ge 2$ and r satisfying $1 < r < R^{1/m}$ be fixed. The rate of convergence is given by

$$\|S_n(f)(z) - f(z)\| \le \frac{2\sqrt{2}\||f|\|_{r^m}}{(r-1)^2} \cdot \frac{1}{r^{(m-1)(n+1)-2}},$$

for all $z \in int\,(E_r) \bigcup E_r$ and $n \in \mathbb{N}$. Here $\||f|\|_r = \sup\{\|f(z)\|; |z| \le r\}$.

Proof. By, e.g., Mason–Handscomb [107], p. 22, relation (2.14), we can write

$$e_j(z) = \sum_{p=0}^{[j/2]} C_{j,j-2p} T_{j-2p}(z) = \sum_{p=0}^{[j/2]} C^*_{j,j-2p} P_{j-2p}(z),$$

where $C^*_{j,j-2p} = 2^{1-j} \binom{j}{p} \frac{\sqrt{\pi}}{\sqrt{2}}$ if $j-2p \ne 0$ and $C^*_{j,j-2p} = 2^{-j} \binom{j}{p} \sqrt{\pi}$ if $j-2p = 0$.
Denoting $j - 2p = q$, we obtain

$$e_j(z) = \sum_{q=j-2[j/2]}^{j} C^*_{j,q} \cdot P_q(z),$$

where $C^*_{j,q} = 2^{1-j} \binom{j}{[(j-q)/2]} \frac{\sqrt{\pi}}{\sqrt{2}}$ if $q \ne 0$ and $C^*_{j,q} = 2^{-j} \binom{j}{[j/2]} \sqrt{\pi}$ if $q = 0$.
Clearly we have $0 < C^*_{j,q} < 2^{1-j} 2^j \sqrt{\pi} = 2\sqrt{\pi}$ for all j and $0 \le q \le j$.
Also, by Mason and Handscomb [107], p. 25, relation (1.50), we have

$$|T_k(z)| \le \frac{r^k + r^{-k}}{2} \le r^k, \text{ for all } z \in E_r.$$

Since the major semiaxis of the ellipse E_r is $a = \frac{r+r^{-1}}{2} \le r < R$, it follows that the interior of the ellipse E_r and its boundary are included in $\mathbb{D}_R$.

By using the estimate in Corollary 3.2.5 and the above considerations, we obtain

$$\|S_n(f)(z) - f(z)\| \le$$

$$\sum_{j=n+1}^{\infty} \|c_j\| \cdot \sum_{q=n+1}^{j} |C^*_{j,q}| \cdot |P_q(z)| \le \sum_{j=n+1}^{\infty} \|c_j\| \sum_{q=n+1}^{j} 2\sqrt{\pi} \frac{\sqrt{2}}{\sqrt{\pi}} r^q =$$

$$2\sqrt{2} \sum_{j=n+1}^{\infty} \|c_j\| \sum_{q=n+1}^{j} r^q = 2\sqrt{2} \sum_{j=n+1}^{\infty} \|c_j\| r^{n+1} \frac{r^{j-n}-1}{r-1} \le$$

$$\frac{2\sqrt{2}}{r-1} \sum_{j=n+1}^{\infty} \|c_j\| r^{n+1} r^{j-n} = \frac{2r\sqrt{2}}{r-1} \sum_{j=n+1}^{\infty} \|c_j\| r^j,$$

for all $z \in \text{int}(E_r) \bigcup E_r$.

(i) Let $1 < r < R$. By $\sum_{j=0}^{\infty} \|c_j\| r^j < \infty$, we have

$$\lim_{n \to \infty} \sum_{j=n+1}^{\infty} \|c_j\| r^j = 0.$$

(ii) Now, let $1 < r < R^{1/m}$. By Cauchy's estimates of the coefficients in the Taylor expansion (see Hille–Phillips [83], p. 97, formula (3.11.3)), we get $\|c_j\| \le \frac{\||f\||_{r^m}}{r^{mj}}$ for all $j \in \mathbb{N} \bigcup \{0\}$, which together with the above estimate in (i) (useful since $r < R^{1/m}$ implies $r < R$) implies

$$\|S_n(f)(z) - f(z)\| \le \frac{2r\sqrt{2}}{(r-1)} \sum_{j=n+1}^{\infty} \||f\||_{r^m} \cdot r^{-(m-1)j} \le$$

$$\frac{2r\sqrt{2}\||f\||_{r^m}}{(r-1)} \cdot \frac{1}{r^{(m-1)(n+1)}} \left[1 + \frac{1}{r} + \dots\right] = \frac{2r^2\sqrt{2}\||f\||_{r^m}}{(r-1)^2} \cdot \frac{1}{r^{(m-1)(n+1)}} =$$

$$\frac{2\sqrt{2}\||f\||_{r^m}}{(r-1)^2} \cdot \frac{1}{r^{(m-1)(n+1)-2}},$$

for all $z \in \text{int}(E_r) \bigcup E_r$ and $n \in \mathbb{N}$, which proves the theorem.   □

Remarks. 1) The order of convergence in Theorem 3.3.1, (i), is still that of a geometric series. Indeed, let $1 < r < r_0 < R$. We obtain

$$\sum_{k=n+1}^{\infty} \|c_k\| r^k = \sum_{k=n+1}^{\infty} \|c_k\| r_0^k \left(\frac{r}{r_0}\right)^k.$$

Now since $\sum_{k=0}^{\infty} \|c_k\| r_0^k < \infty$, it follows that $\lim_{k \to \infty} \|c_k\| r_0^k = 0$, which implies that there exists a constant $M(r_0) > 0$ such that

$$\|c_k\| r_0^k \le M(r_0), \text{ for all } k \in \mathbb{N}.$$

Therefore, denoting $\rho = \frac{r}{r_0} < 1$, we get

$$\sum_{k=n+1}^{\infty} \|c_k\| r^k \le M(r_0) \sum_{k=n+1}^{\infty} \rho_0^k = \frac{M(r_0)}{1 - \rho_0} \rho_0^{n+1}.$$

2) For example, by taking $m = 3$ and $X = \mathbb{C}$ in Theorem 3.3.1, (ii), it follows that the order of approximation on $[-1, 1]$ is $O(1/r^{2n})$, which is essentially better than the order $O(1/r^n)$ in Theorem 5.16, p. 130 in Mason and Handscomb [107] already mentioned at the beginning of this section.

3) If, in addition, in the statement of Theorem 3.3.1, we suppose that f is continuous in $\overline{\mathbb{D}}_R$, then it follows that $\||f\||_{r^m} \le \||f\||_R$, and therefore if $1 < r \le R^{1/m}$ then for all $z \in \text{int}(E_r) \bigcup E_r$ and $n, m \in \mathbb{N}$, we have

$$\|S_n(f)(z) - f(z)\| \leq \frac{2\sqrt{2}\|\|f\|\|_R}{(r-1)^2} \cdot \frac{1}{r^{(m-1)(n+1)-2}}.$$

Unfortunately, by fixing $n \in \mathbb{N}$ and passing with $m \to \infty$ in the previous estimate (which geometrically means that $r \to 1$ and the ellipse E_r degenerates to the interval $[-1,1]$), the right-hand side of the above inequality does not converge to zero. Indeed, by taking, for example, $r = R^{1/m}$, we obtain

$$\frac{2\sqrt{2}\|\|f\|\|_R}{(r-1)^2} \cdot \frac{1}{r^{(m-1)(n+1)-2}} \geq \frac{2\sqrt{2}\|\|f\|\|_R}{r^2} \cdot \frac{1}{r^{(m-1)(n+1)-2}} = \frac{2\sqrt{2}\|\|f\|\|_R}{r^{(m-1)(n+1)}} =$$

$$\frac{2\sqrt{2}\|\|f\|\|_R}{R^{(m-1)(n+1)/m}} \geq \frac{2\sqrt{2}\|\|f\|\|_R}{R^{n+1}}.$$

4) In the case when $X = \mathbb{C}$, a lower estimate in approximation of $f(z)$ by $S_n(f)(z)$ can easily be derived from the general Theorem 4, p. 130 in Cheney [25] on orthogonal expansions, or better from the particular result for Chebyshev expansions in Theorem 5, p. 131 in Cheney [25]. In this sense, we immediately obtain

$$\sup_{z \in \text{int}(E_r) \cup E_r} |S_n(f)(z) - f(z)| \geq \frac{\pi}{4} \max\{|a_{n+1}(f)|, |a_{n+2}(f)|, \dots, \},$$

for all $n \in \mathbb{N}$, where $a_k(f) = < f, P_k >_\rho$.

In what follows, similar results for the Legendre polynomials will be obtained. Before going into details concerning these orthogonal polynomials, we need to prove a very useful elementary result.

Lemma 3.3.2. Let $r > 1$ and $z \in \mathbb{C}$. If $|z + \sqrt{z^2 - 1}| = r$, then for all $\lambda \in \mathbb{R}$ satisfying $|\lambda| \leq 1$, we have $|z + \lambda\sqrt{z^2 - 1}| \leq r$.

Proof. We have two possible cases: 1) $\lambda \geq 0$; 2) $\lambda < 0$.

Case 1. We get

$$|z + \lambda\sqrt{z^2 - 1}| = |\lambda[z + \sqrt{z^2 - 1}] + (1 - \lambda)z| \leq \lambda r + (1 - \lambda)|z|.$$

But when z satisfies $|z + \sqrt{z^2 - 1}| = r$, by, e.g., Mason and Handscomb [107], p. 12–14, relations (1.38), (1.39), and (1.44), it follows that z is on the ellipse with foci at -1 and 1 and with the semiaxis $a = \frac{1}{2}\left(r + \frac{1}{r}\right)$ and $b = \frac{1}{2}\left(r - \frac{1}{r}\right)$. Therefore the maximum of $|z|$ is attained evidently for $z = a$, which replaced in the last inequality immediately implies

$$|z + \lambda\sqrt{z^2 - 1}| \leq \lambda r + (1 - \lambda)\frac{1}{2}\left(r + \frac{1}{r}\right) \leq r.$$

Case 2. Since $[z + \sqrt{z^2 - 1}][z - \sqrt{z^2 - 1}] = 1$, by $|z + \sqrt{z^2 - 1}| = r$, we immediately obtain that $|z - \sqrt{z^2 - 1}| = 1/r$.

Now denote $\mu = -\lambda \in [0, 1]$. Reasoning as in Case 1, we immediately obtain

$$|z + \lambda\sqrt{z^2 - 1}| = |z - \mu\sqrt{z^2 - 1}| = |\mu[z - \sqrt{z^2 - 1}] + (1 - \mu)z| \leq$$

$$\mu \cdot \frac{1}{r} + (1 - \mu)|z| \leq \frac{\mu}{r} + (1 - \mu)\frac{1}{2}\left(r + \frac{1}{r}\right) \leq r,$$

which proves the lemma. □.

Concerning quantitative estimates for the approximation by the Fourier-type sums with respect to the sequence of classical Legendre polynomials given by

$$L_k(z) = \frac{1}{2^k k!}\frac{d^k}{dx^k}[(x^2 - 1)^k], \ k = 0, 1, 2, \ldots,$$

we present the following result.

Theorem 3.3.3. *Let* $(X, \|\cdot\|)$ *be a complex Banach space. Suppose that* $f :$ $\mathbb{D}_R \to X$ *is analytic in* $\mathbb{D}_R$ *with* $R = \sqrt{2}$*, that is,* $f(z) = \sum_{j=0}^{\infty} c_j z^j$ *for all* $z \in \mathbb{D}_R$*, and there exists* $A \in (0, 1)$ *and* $M > 0$ *such that* $\|c_j\| \leq M\frac{A^j}{j!}$*, for all* j *(this evidently implies that* $\|f(z)\| \leq Me^{A|z|}$*, i.e., that* f *is of exponential growth).*
Define

$$S_n(f)(z) = \sum_{k=0}^{n} < f, P_k >_{\rho,X} \cdot P_k(z),$$

the partial sum of the Legendre series attached to f*, where*

$$P_k(x) = \frac{\sqrt{2k + 1}}{\sqrt{2}}L_k(x), \ k = 0, 1, 2, \ldots,$$

denotes the sequence of orthonormalized Legendre polynomials on $[-1, 1]$ *with respect the weight* $\rho(x) = 1$*. Also, denote by* E_r *the ellipse with the semiaxis* $a = \frac{r + r^{-1}}{2}$ *and* $b = \frac{r - r^{-1}}{2}$ *(where* $r > 1$*), with foci at* -1 *and* 1*.*
For any r *satisfying* $1 < r < \sqrt{2}$*, the rate of convergence is given by*

$$\|S_n(f)(z) - f(z)\| \leq \frac{4r^2}{(2 - r^2)(1 - A)}A^{n+1},$$

for all $z \in int \, (E_r) \bigcup E_r$ *and* $n \in \mathbb{N}$*.*

Proof. (i) By, e.g., Mocica [111], p. 225, Problem 4.60, with solution at p. 282, we can write

$$e_j(z) = \sum_{q=j,j-2,\ldots,} C_{j,q}^* \frac{\sqrt{2}}{\sqrt{2q+1}} P_q(z),$$

where

$$C_{j,q}^* = \frac{(2q+1)j!}{2^{(j-q)/2}[(j-q)/2]!(j+q+1)!!}.$$

Here we have taken into account that since $j-q$ in the above sum is always an even number, it follows that $j+q+1$ is odd, that is, $j+q+1 = 2s+1$ with $s = (j+q)/2$, the case when for the double factorial, we have the formula

$$(j+q+1)!! = \frac{(j+q+1)!}{2^s s!}.$$

Now we will prove the estimate

$$C_{j,q}^* \leq \frac{j^{s+1}}{2^{j-2}}, \text{ for all } q = j - 2s, \quad j \in \mathbb{N}, j \geq 2, \quad s = 0, 1, 2, \ldots,.$$

First notice that this is obviously equivalent to

$$C_{j,q}^* \leq \frac{j^{s+1}}{2^{j-2}},$$

for any arbitrary fixed $q \in \mathbb{N}$, $q \geq 2$ and all $j = q + 2s$ with $s = 0, 1, 2, \ldots,$. By simple calculation we get the recurrence formula

$$C_{j+2,q}^* = C_{j,q}^* \frac{(j+1)(j+2)}{(j+2-q)(j+3+q)}.$$

Also, from the formula of definition it is easy to see that

$$C_{q,q}^* = (2q+1)\Pi_{k=1}^q \frac{k}{2k+1} \leq \frac{2q+1}{2^q},$$

which implies

$$C_{q,q}^* \leq \frac{4q}{2^q} \leq \frac{q}{2^{q-2}}.$$

We will reason step by step. Taking in the recurrence formula $j = q$ (i.e., above $s = 0$), we obtain

$$C_{q+2,q}^* = C_{q,q}^* \frac{(q+1)(q+2)}{2(2q+3)} \leq \frac{2q+1}{2^q} \cdot \frac{q+1}{2}.$$

Denoting $q + 2 = j$, the above inequality easily becomes

$$C_{j,j-2}^* \leq \frac{2(j-2)+1}{2^{j-2}} \cdot \frac{j-1}{2} \leq \frac{j^2}{2^{j-2}}.$$

Taking in the recurrence formula $j = q + 2$ (i.e., above $s = 1$), we easily obtain

$$C^*_{q+4,q} \leq C^*_{q+2,q} \frac{q+3}{4} \leq \frac{2q+1}{2^q} \cdot \frac{q+1}{2} \cdot \frac{q+3}{4},$$

where by denoting $q + 4 = j$ and replacing in the previous inequality, we get

$$C^*_{j,j-4} \leq \frac{j^3}{2^{j-2}}.$$

Step by step, in general we easily arrive at

$$C^*_{j,j-2s} \leq \frac{j^{s+1}}{2^{j-2}}, \quad s = 0, 1, 2, \ldots,.$$

Denoting $j - 2s = q$, we get the desired estimate

$$C^*_{j,q} \leq \frac{j^{(j-q)/2+1}}{2^{j-2}}.$$

On the other hand, by, e.g., Lebedev [94], p. 48, formula (4.4.1), we can write

$$L_k(z) = \frac{1}{\pi} \int_0^\pi (z + \sqrt{z^2 - 1}\cos t)^k dt, \quad \text{for all } z \in \mathbb{C},$$

which together with Lemma 3.3.2 (and its proof, Case 1) immediately implies

$$|L_k(z)| \leq r^k, \quad \text{for all } z \in E_r.$$

Since the major semiaxis of the ellipse E_r is $a = \frac{r+r^{-1}}{2} \leq r < R$, it follows that the interior of the ellipse E_r and its boundary are included in $\mathbb{D}_R$.

By using the estimate in Corollary 3.2.5 and the above considerations, for all $z \in \text{int}(E_r) \bigcup E_r$, we obtain

$$\|S_n(f)(z) - f(z)\| \leq$$

$$\sum_{j=n+1}^\infty \|c_j\| \cdot \sum_{q=n+1}^j |C^*_{j,q}| \cdot |L_q(z)| \leq \sum_{j=n+1}^\infty \|c_j\| \sum_{q=n+1}^j \frac{j^{(j-q)/2+1}}{2^{j-2}} \cdot r^q.$$

But since $j/r^2 > 1$ for all $j \geq n + 1 \geq 2$, $n \in \mathbb{N}$ and since in this case we easily get (by mathematical induction)

$$\frac{1}{j - r^2} \leq \frac{j}{2 - r^2} \quad \text{and} \quad j^j \leq 2^j j!,$$

it follows

$$\sum_{q=n+1}^j \frac{j^{(j-q)/2+1}}{2^{j-2}} \cdot r^q \leq \frac{r^j}{2^{j-2}} \sum_{s=0}^{[(j-n-1)/2]+1} \frac{j^{s+1}}{r^{2s}} = \frac{jr^j}{2^{j-2}} \sum_{s=0}^{[(j-n+1)/2]} \left(\frac{j}{r^2}\right)^s \leq$$

$$\frac{jr^j}{2^{j-2}} \sum_{s=0}^{[j/2]} \left(\frac{j}{r^2}\right)^s = \frac{jr^j}{2^{j-2}} \cdot \frac{r^2}{j-r^2} \cdot [(j/r^2)^{[j/2]+1}-1] \leq \frac{1}{2-r^2} \cdot \frac{j^{[j/2]+2}r^{j-2[j/2]}}{2^{j-2}}$$

$$\leq \frac{r^2}{2-r^2} \cdot \frac{j^{[j/2]+2}}{2^{j-2}} \leq \frac{4r^2}{2-r^2} \cdot \left(\frac{j}{2}\right)^j \leq \frac{4r^2}{2-r^2} \cdot j!,$$

where $1 < r^2 < 2$. In conclusion,

$$\|S_n(f)(z) - f(z)\| \leq \frac{4r^2}{2-r^2} \sum_{j=n+1}^{\infty} \|c_j\| j! \leq \frac{4Mr^2}{2-r^2} \sum_{j=n+1}^{\infty} A^j =$$

$$\frac{4Mr^2}{(2-r^2)(1-A)} A^{n+1},$$

which proves the theorem.   □

Remark. According to, e.g., Theorem 12.4.7 and Lemma 4.4.2 (and their proofs) in Davis [30], it follows that if the complex-valued function f is analytic in the interior of the ellipse E_r, then $S_n(f)(z)$ converges uniformly to $f(z)$ in any closed set included in E_r, with the rate of a geometric series. But the geometric series and the constants are not explicitly given there. Under the stronger hypothesis that f is of exponential growth, the above Theorem 3.3.3 in the particular case when $X = \mathbb{C}$ gives the estimate with explicit constant and order of convergence.

3.4 Notes and Open Problems

Note 3.4.1. For the particular case $X = \mathbb{C}$, all the results in this chapter were obtained in Anastassiou and Gal [10].

Note 3.4.2. We recall some old results obtained by Jackson [85, 86] concerning the overconvergence phenomenon for the Taylor expansion.

Let $x_{n,k} \in (a,b), k = 1, \ldots, q_n$, with $q_n \leq n$, where (a,b) is a finite subinterval of $\mathbb{R}$ and $\mu_{n,k}, k = 1, \ldots, q_n$ are positive integers such that $\sum_{k=1}^{q_n} \mu_{n,k} = p_n < n$. Also, let $P_n(x)$ be a polynomial of degree n with real coefficients, satisfying the Taylor interpolation conditions

$$P_n^{(j)}(x_{n,k}) = f^{(j)}(x_{n,k}), \text{ for all } j = 0, 1, \ldots, \mu_{n,k}, k = 1, \ldots, q_n.$$

Since $p_n - 1 < n$, there are infinitely many polynomials $P_n(x)$ satisfying the above conditions. In particular, we are looking for that polynomial $P_n(x)$ which makes minimum the value of the integral $\int_a^b [f(x) - P_n(x)]^2 dx$.

Theorem 3.4.1 (Jackson [85]). *If f is analytic throughout the interior and on the boundary of a circle of the complex plane with center on the middle of*

the segment $[a, b]$ and with radius $R > 2r$, where $r = (b-a)/2$, and if $f(x)$
is real for x real, then the polynomial $P_n(x)$ minimizing the above integral
converges toward $f(x)$ for all x throughout the interior and on the boundary
of any circle of radius $\rho < R - 2r$, concentric with the circle of analyticity
of f.

Remark. The result in Theorem 3.4.1 was extended in Jackson [86] to the
case of minimizing the more general integral $\int_a^b \rho(x)[f(x) - P_n(x)]^m dx$, where
$m > 0$ and $\rho(x)$ is a weight with the property $\int_a^b \rho(x)dx > 0$.

Note 3.4.3. Theorem 3.2.4, Corollary 3.2.5, Theorem 3.3.1, and Theorem
3.3.3 appear for the first time here.

Note 3.4.4. It is worth noting that in the case of trigonometric Fourier expansion, if we suppose in addition the analyticity of f in a strip containing
the real axis, then that improves the convergence on $\mathbb{R}$ (see de la Vallée–
Poussin [141], pp. 110–150, namely, Chaps. 8–10). More exactly, as a simple
sample, recall here the following result (see Vallée–Poussin [141], p. 111, Theorem II): *if f is holomorphic (analytic) and bounded by $M > 0$ in the strip*
$S_b = \{z = x + iy \in \mathbb{C}; x \in \mathbb{R}, |y| \le b\}$, *then the order of approximation on*
the real axis of the Fourier partial sum of order n is $\le \frac{2M}{e^b - 1} e^{-nb}$.

Open Problem 3.4.5. It is an open question to apply the general result Corollary 3.2.5 to the system of Hermite polynomials, orthogonal on
$(-\infty, +\infty)$ with respect to the weight $\rho(x) = e^{-x^2}$, that is,

$$\int_{-\infty}^{+\infty} e^{-x^2} H_n(x)H_m(x)dx = 0, \text{ for } n \ne m,$$

and

$$\int_{-\infty}^{+\infty} e^{-x^2} H_n(x)H_m(x)dx = 2^n n! \sqrt{\pi}, \text{ for } n = m.$$

The following auxiliary result could be useful.

Lemma A. *For all $z = x + iy$ with $x \in \mathbb{R}$ and $|y| \le d$ (i.e., z is in a strip*
around the OX-axis) and $k = 0, 1, 2, \ldots$, we have the estimates

$$|\sin(kz)| \le e^{k|y|} \le e^{kd}, \quad |\cos(kz)| \le e^{k|y|} \le e^{kd}.$$

Proof. Taking into account that $\sinh(y) = (e^y - e^{-y})/2$ for $y \in \mathbb{R}$, we get

$$|\sin(z)| = \sqrt{\sin^2(x) + \sinh^2(y)} = \sqrt{(e^{2y} + e^{-2y})/2 - \cos^2(x)} \le e^{|y|}.$$

Replacing now z by kz, we obtain the required inequality in the statement
for sin.

For the upper estimate of $|\cos(z)|$, we take into account that

$$|\cos(z)| = |\sin(\pi/2 - z)| = |\sin(\pi/2 - x - iy)| = \sqrt{\cos^2(x) + \sinh^2(y)} =$$

$$\sqrt{(e^{2y} + e^{-2y})/2 - \sin^2(x)} \le e^{|y|},$$

which proves the estimate for $|\cos(kz)|$ too. □.

Also, the following representation formulas (see, e.g., Mocica [111], p. 238, formula in Problem 4.112, c))

$$H_k(z) = \frac{2^{k+1}}{\sqrt{\pi}} e^{z^2} \int_0^\infty e^{-t^2} t^k \cos\left(2zt - \frac{\pi k}{2}\right) dt, z \in \mathbb{C}$$

and (see, e.g., Rainville [125], p. 110, formula (4))

$$z^j = \frac{j!}{2^j} \sum_{k=0}^{[j/2]} \frac{1}{k!(j-2k)!} H_{j-2k}(z)$$

could be useful.

Open Problem 3.4.6. It is an open question to apply the general result Corollary 3.2.5 to the Laguerre polynomials $(L_n^{(\lambda)})_{n \in \mathbb{N} \cup \{0\}}$, $\lambda > -1$, orthogonal on $(0, +\infty)$ with respect to the weight $\rho(x) = x^\lambda e^{-x}$, that is,

$$\int_{-\infty}^{+\infty} x^\lambda e^{-x} L_n^{(\lambda)}(x) L_m^{(\lambda)}(x) dx = 0, \text{ for } n \ne m,$$

and

$$\int_{-\infty}^{+\infty} x^\lambda e^{-x} L_n^{(\lambda)}(x) L_m^{(\lambda)}(x) dx = \frac{\Gamma(\lambda + n + 1)}{n!}, \text{ for } n = m.$$

Here $\Gamma(u) = \int_0^\infty e^{-t} t^{u-1} dt, u > 0$ denotes the Gamma function.

We have the integral representation (see, e.g., Mocica [111], p. 245, formula (4.63))

$$L_n^{(\lambda)}(z) = \frac{1}{n!} e^z e^{-\lambda/2} \int_0^\infty t^{n+\lambda/2} e^{-t} J_\lambda(2\sqrt{tz}) dt,$$

where $\lambda > -1$, $z \in \mathbb{C}$ and $J_\lambda(u)$ is the Bessel function of the first kind which by, e.g., Mocica [111], p. 150, formula (3.25), has the integral representation

$$J_\lambda(z) = \frac{1}{\sqrt{\pi}\Gamma(\lambda + 1/2)} \left(\frac{z}{2}\right)^\lambda \int_{-1}^{+1} (1 - u^2)^{\lambda - 1/2} \cos(uz) du, \lambda > -1/2.$$

Also, we can write (see, e.g., Rainville [125], p. 118, formula (2))

$$z^j = (1 + \lambda)_j \sum_{k=0}^{j} \frac{(-j)_k}{(1 + \lambda)_k} L_k^{(\lambda)}(z),$$

where $(\lambda)_p = \lambda(\lambda + 1) \ldots (\lambda + p - 1)$.

All these formulas could be useful in solving the open problem.

Open Problem 3.4.7. It is an open question to apply the general result Corollary 3.2.5 to the Gegenbauer (or ultraspherical) polynomials that are given by (see, e.g., Szegö [138], p. 80)

$$G_k^{(\lambda)}(z) = \frac{\Gamma(\lambda + 1/2)}{\Gamma(2\lambda)} \cdot \frac{\Gamma(k + 2\lambda)}{\Gamma(k + \lambda + 1/2)} J_k^{(\alpha,\beta)}(z), z \in \mathbb{C},$$

where $J_k^{(\alpha,\beta)}(z)$ denotes the Jacobi polynomials and $\alpha = \beta = \lambda - 1/2, \lambda > -1/2$.

They are orthogonal polynomials on $[-1, 1]$ with respect to the weight $\rho(x) = (1 - x^2)^{\lambda - 1/2}$, and we have

$$\int_{-1}^{1} \rho(x)[G_k^{(\lambda)}]^2 dx = 2^{1-2\lambda} \pi \frac{\Gamma(k + 2\lambda)}{(k + \lambda)k!\Gamma^2(\lambda)}.$$

In solving the open problem, the integral representation (see, e.g., Abramovitz [4], p. 784, formula **22.10.10**)

$$G_k^\lambda(z) = \frac{2^{1-2\lambda}\Gamma(k + 2\lambda)}{k!(\Gamma(\lambda))^2} \int_0^\pi (z + \sqrt{z^2 - 1}\cos(t))^k \sin^{2\lambda-1}(t)dt, z \in \mathbb{C},$$

and the formula (see, e.g., Rainville [125], p. 144, formula (36))

$$e_j(z) = \frac{j!}{2^j} \sum_{k=0}^{[j/2]} \frac{j + \lambda - 2k}{k!(\lambda)_{j+1-k}} G_{j-2k}^\lambda(z),$$

valid for all $j \in \mathbb{N} \bigcup \{0\}$, could be useful, where $(\lambda)_p = \lambda(\lambda+1) \ldots (\lambda+p-1)$. Also, the above Lemma A in Open Problem 3.4.5 could be of interest here.

References

1. Abel, U., Gupta, V., Mohapatra, R.N.: Local approximation by beta operators. Nonlinear Anal. **62**(1), 41–52 (2005)
2. Abel, U., Gupta, V., Mohapatra, R.: Local approximation by a variant of Bernstein-Durrmeyer operators. Nonlinear Anal. Theory Methods Appl. **68**(11), 3372–3381 (2008)
3. Abel, U., Heilmann, H.: The complete asymptotic expansion for Bernstein-Durrmeyer operator with Jacobi weights. Mediterr. J. Math. **1**(4), 487–499 (2004)
4. Abramowitz, M., Stegun, I.A.: Handbook of Mathematical Functions. Dover Publications, New York (1964)
5. Alexander, J.W.: Functions which map the interior of the unit circle upon simple regions. Ann. Math. **17**(1915), 12–22 (1915)
6. Aliev, I.A., Gadjiev, A.D., Aral, A.: On approximation properties of a family of linear operators at critical value of parameter. J. Approx. Theory **138**, 242–253 (2006)
7. Altomare, F., Mangino, E.: On a generalization of Baskakov operator. Rev. Roumaine Math. Pures Appl. **44**(5–6), 683–705 (1999)
8. Altomare, F., Raşa, I.: Feller semigroups, Bernstein-type operators and generalized convexity associated with positive projections. New Develpments in Aproximation Theory, Dortmund, 1998, pp. 9–32. Birkhauser, Basel 1999 (1999)
9. Anastassiou, G., Gal, S.G.: Approximation by complex Bernstein-Schurer and Kantorovich-Schurer polynomials in compact disks. Comput. Math. Appl. **58**(4), 734–743 (2009)
10. Anastassiou, G., Gal, S.G.: Quantitative estimates in the overconvergence of Chebyshev and Legendre orthogonal expansions on [−1,1]. Nonlinear Stud. **2**, 139–149 (2010)
11. Andrews, G.E., Askey, R., Roy, R.: Special Functions. Cambridge University Press, Cambridge (2000)
12. Bartolomeu, J., He, M.: On Faber polynomials generated by an m-star. Math. Comput. **62**, 277–288 (1994)
13. Beatson, R.K.: Bell-shape preserving convolution operators. Technical Report, Department of Mathematics, University of Otago, Dunedin, New Zealand, 1978
14. Beise, P., Meyrath, T., Müller, J.: Universality properties of Taylor series inside the domain of holomorphy. J. Math. Anal. Appl. **383**(1), 234–238 (2011)
15. Berens, H., Xu, Y.: On Bernstein-Durrmeyer polynomials with Jacobi weights. In: Chui, C.K. (ed.) Approximation Theory and Functionals Analysis, pp. 25–46. Academic Press, Boston (1991)
16. Bernstein, S.N.: Sur la convergence de certaines suites des polynômes. J. Math. Pures Appl. **15**(9), 345–358 (1935)

17. Bernstein, S.N.: Sur le domaine de convergence des polynômes. C.R. Acad. Sci. Paris **202**, 1356–1358 (1936)
18. Bernstein, S.N.: On the domains of convergence of polynomials (in Russian). Izv. Akad. Nauk. SSSR Ser. Math. **7**, 49–88 (1943)
19. Bernstein, S.N.: Complétement a l'article de E. Voronowskaja. C.R. Acad. Sci. U.R.S.S. Ser. A. **4**, 86–92 (1932)
20. Bernstein, S.N.: Leçns sur les Propriétés Extrémales et la Meilleure Approximations des Fonctions Analytiques d'Une Variable Réelle. Gauthier-Villars, Paris (1926)
21. Bleimann, G, Butzer, P.L., Hahn, L.: A Bernstein-type operator approximating continuous functions on the semi-axis. Indag. Math. **42**, 255–262 (1980)
22. Bourion, G.: L' Ultracovergence dans les Séries de Taylor, Actualités Scientifiques et Industrielle, No. **472**, Paris (1937)
23. Boyd, J.B.: Chebyshev and Fourier Spectral Methods, 2nd edn. Dover Publications, Mineola, New York (2000)
24. Chen, W.Z.: On the modified Bernstein-Durrmeyer operators. In: Report of the Fifth Chinese Conference on Approximation Theory, Zhen Zhou, China (1987)
25. Cheney, E.W.: Introduction to Approximation Theory. McGraw-Hill, New York (1966)
26. Cimoca, G., Lupaş, A.: Two generalizations of the Meyer-König and Zeller operator. Mathematica(Cluj) **9(32)**(2), 233–240 (1967)
27. Coleman, J.P., Smith, R.A.: The Faber polynomials for circular sectors. Math. Comput. **49**, 231–241 (1987)
28. Coleman, J.P., Myers, N.J.: The Faber polynomials for annular sectors. Math. Comput. **64**, 181–203 (1995)
29. Curtiss, J.H.: Faber polynomials and the Faber series. Am. Math. Month **78**(6), 5677–596 (1971)
30. Davis, Ph.J.: Interpolation and Approximation. Blaisdell Publishing Company, Waltham-Massachussetts-Toronto-London (1963)
31. DeVore, R.A., Lorentz, G.G.: Constructive Approximation: Polynomials and Splines Approximation, vol. 303. Springer, Berlin, Heidelberg (1993)
32. Dressel, F.G., Gergen, J.J., Purcell, W.H.: Convergence of extended Bernstein polynomials in the complex plane. Pacific J. Math. **13**(4), 1171–1180 (1963)
33. Dzjadyk, V.K.: Introduction to the Theory of Uniform Approximation of Functions by Polynomials (Russian). Nauka, Moscow (1977)
34. Faber, G.: Über polynomische Entwicklungen. Math. Ann. **57**, 398–408 (1903)
35. Flett, T.M.: Temperatures, Bessel potentials and Lipschitz space. Proc. Lond. Math. Soc. **22**(3), 385–451 (1971)
36. Gadjiev, A.D., Aral, A., Aliev, I.A.: On behaviour of Riesz and generalized Riesz potentials as order tends to zero. Math. Inequalities Appl. **4**, 875–888 (2007)
37. Gaier, D.: Approximation durch Fejér-Mittel in der Klausse A. Mitt. Math. Sem. Giessen **123**, 1–6 (1977)
38. Gaier, D.: Lectures on Complex Approximation. Birkhauser, Boston (1987)
39. Gal, S.G.: Shape Preserving Approximation by Real and Complex Polynomials. Birkhauser, Boston, Basel, Berlin (2008)
40. Gal, S.G.: On Beatson convolution operators in the unit disk. J. Anal. **10**, 101–106 (2002)
41. Gal, S.G.: Voronovskaja's theorem and iterations for complex Bernstein polynomials in compact disks. Mediterr. J. Math. **5**(3), 253–272 (2008)
42. Gal, S.G.: Geometric and approximate properties of convolution polynomials in the unit disk. Bull. Inst. Math. Acad. Sinica (New Series) **1**, 307–336 (2006)
43. Gal, S.G.: Exact orders in simultaneous approximation by complex Bernstein polynomials. J. Concr. Appl. Math. **7**(3), 215–220 (2009)
44. Gal, S.G.: Approximation by complex Bernstein-Stancu polynomials in compact disks. Results Math. **53**(3–4), 245–256 (2009)

45. Gal, S.G.: Exact orders in simultaneous approximation by complex Bernstein-Stancu polynomials. Revue d'Anal. Numér. Théor. de L'Approx. (Cluj-Napoca) **37**(1), 47–52 (2008)

46. Gal, S.G.: Generalized Voronovskaja's theorem and approximation by Butzer's combinations of complex Bernstein polynomials. Results Math. **53**(3–4), 257–268 (2009)

47. Gal, S.G.: Approximation by complex Bernstein-Kantorovich and Stancu-Kantorovich polynomials and their iterates in compact disks. Revue D'Anal. Numér. Théor. de L'Approx. (Cluj) **37**(2), 159–168 (2008)

48. Gal, S.G.: Degree of approximation of continuous functions by some singular integrals. Revue D'Analyse Numér. Théor. Approx. (Cluj) **27**(2), 251–261 (1998)

49. Gal, S.G.: Approximation by Complex Bernstein and Convolution Type Operators. World Scientific Publishing, New Jersey, London, Singapore, Beijing, Shanghai, Hong Kong, Taipei, Chennai (2009)

50. Gal, S.G.: Approximation by complex genuine Durrmeyer type polynomials in compact disks. Appl. Math. Comput. **217**, 1913–1920 (2010)

51. Gal, S.G.: Approximation by complex Bernstein-Durrmeyer polynomials with Jacobi weightds in compact disks. Mathematica Balkanica (N.S.) **24**(1–2), 103–119 (2010)

52. Gal, S.G.: Approximation by complex Lorentz polynomials. Math. Comm. **16**(2011), 65–75 (2011)

53. Gal, S.G.: Differentiated generalized Voronovskaja's theorem in compact disks. Results Math. **61**(3), 247–253 (2012)

54. Gal, S.G.: Tchebycheff orthogonal expansions for vector-valued functions. Anal. Univ. Oradea Fasc. Matem. **11**, 111–118 (2004)

55. Gal, S.G.: Convolution type integral operators in complex approximation. Comput. Meth. Funct. Theory **1**(2), 417–432 (2001)

56. Gal, S.G.: Approximation by complex potentials generated by the Gamma function. Turkish J. Math. **35**(3), 443–456 (2010)

57. Gal, S.G.: Approximation by complex potentials generated by the Euler's Beta function. Eur. J. Pure Appl. Math. **3**(6), 1150–1164 (2010)

58. Gal, S.G.: Approximation by complex q-Lorentz polynomials, $q > 1$. Mathematica (Cluj) **54(77)**(1), 53–63 (2012)

59. Gal, S.G.: Approximation of analytic functions without exponential growth conditions by complex Favard-Szász-Mirakjan operators. Rendiconti del Circolo Matematico di Palermo **59**(3), 367–376 (2010)

60. Gal, S.G.: Approximation by quaternion q-Bernstein polynomials, $q > 1$. Adv. Appl. Clifford Alg. **22**(2), 313–319 (2012)

61. Gal, S.G.: Approximation in compact sets by q-Stancu-Faber polynomials, $q > 1$. Comput. Math. Appl. **61**(10), 3003–3009 (2011)

62. Gal, S.G.: Voronovskaja-type results in compact disks for quaternion q-Bernstein operators, $q \geq 1$. Complex Anal. Oper. Theory **6**(2), 515–527 (2012)

63. Gal, S.G.: (Online access) Erratum to: Differentiated generalized Voronovskaja's theorem in compact disks. Results Math. DOI 10.1007/s00025-012-0295-1, published online 23 October 2012

64. Gal, G.C., Gal, S.G., Goldstein, J.A.: Higher order heat and Laplace type equations with real time variable and complex spatial variable. Complex Variables Elliptic Equat. **55**(4), 357–373 (2010)

65. Gal, S.G., Greiner, R.: Convolution polynomials through Beatson kernels in the unit disk. Anal. Univ. Oradea Fasc. Math. **XVII**(2), 213–217 (2010)

66. Gal, S.G., Gupta, V.: (Online access) Approximation by complex Beta operators of first kind in strips of compact disks. Mediterranean J. Math. **10**(1), 31–39 (2013)

67. Gal, S.G., Gupta, V., Mahmudov, N.I.: Approximation by a complex q-Durrmeyer type operator. Ann. Univ. Ferrara **58**, 65–87 (2012)

68. Gal, S.G., Mahmudov, N.I., Kara, M.: (Online access) Approximation by complex q-Szàsz-Kantorovich operators in compact disks. Complex Anal. Oper. Theory, DOI: 10.1007/s11785-012-0257-3

69. Gentili, G., Stoppato, C.: Power series and analyticity over the quaternions. Math. Ann. **352**(1), 113–131 (2012)

70. Gentili, G., Struppa, D.C.: A new theory of regular functions of a quaternionic variable. Advances Math. **216**, 279–301 (2007)

71. Gonska, H., Piţul, P., Raşa, I.: On Peano's form of the Taylor remainder, Voronovskaja's theorem and the commutator of positive linear operators. In: Proceed. Intern. Conf. on "Numer. Anal., Approx. Theory", NAAT, Cluj-Napoca, Casa Cartii de Stiinta, Cluj-Napoca, pp. 55–80, 2006

72. Gonska H., Raşa, I.: Asymptotic behaviour of differentiated Bernstein polynomials. Mat. Vesnik, **61**, 53–60 (2009)

73. Goodman, T.N.T., Sharma, A.: A modified Bernstein-Schoenberg operator. In: Sendov, Bl. et al (eds.) Constructive Theory of Functions - Varna 1987, pp. 166–173. Bulgar. Acad. Sci., Sofia (1988)

74. Graham, I., Kohr, G.: Geometric Function Theory in One and Higher Dimensions, Pure and Applied Mathematics, vol. 255. Marcel Dekker, New York (2003)

75. Gupta, V.: Some approximation properties of q-Durrmeyer type operators. Appl. Math. Comput. **197**(1), 172–178 (2008)

76. Gupta, V., Finta, Z.: On certain q-Durrmeyer type operators. Appl. Math. Comput. **209**(2), 415–420 (2009)

77. Hasson, M.: Expansion of analytic functions of an operator in series of Faber polynomials. Bull. Aust. Math. Soc. **56**, 303–318 (1997)

78. He, M.: Explicit representations of Faber polynomials for m-cusped hypocycloids. J. Approx. Theory **87**, 137–147 (1996)

79. He, M.: The Faber polynomials for m-fold symmetric domains. J. Comput. Appl. Math. **54**, 313–324 (1994)

80. He, M.: The Faber polynomials for circular lunes. Comput. Math. Appl. **30**, 307–315 (1995)

81. Henrici, P.: Applied and Computational Analysis, vol. I. Wiley, New York (1974)

82. He, M., Saff, E.B.: The zeros of Faber polynomials for and m-cusped hypocycloid. J. Approx. Theory **78**, 410–432 (1994)

83. Hille, E., Phillips, R.S.: Functional Analysis and Semigroups, vol. **31**. American Mathematical Society Colloquium Publications, Providence, RI (1957)

84. Ilieff, L.: Analytische Nichtfortsetzbarkeit und Überkonvergenz einiger Klassen von Potenzriehen, Mathematische Forchungsberichte, vol. XII. VEB Deutscher Verlag der Wissenschaften, Berlin (1960)

85. Jackson, D.: On the approximate representation of analytic functions. Bull. Am. Math. Soc. **34**, 56–62 (1928)

86. Jackson, D.: Note on the convergence of a sequence of approximating polynomials. Bull. Am. Math. Soc. **37**, 69–72 (1931)

87. Jackson, F.H.: On q-definite integrals. Q. J. Pure Appl. Math. **41**, 193–203 (1910)

88. Jakimovski, A., Sharma, A., Szabados, J.: Walsh Equiconvergence of Complex Interpolating Polynomials, Springer Monographs in Mathematics. Springer, Dordrecht (2006)

89. Kantorovitch, L.V.: Sur la convergence de la suite de polynômes de S. Bernstein en dehors de l'interval fundamental. Bull. Acad. Sci. URSS **8**, 1103–1115 (1931)

90. Khan, M.K.: Approximation properties of Beta operators. In: Progress in Approximation Theory, pp. 483–495. Academic Press, New York (1991)

91. Kohr, G., Mocanu, P.T.: Special Chapters of Complex Analysis (in Romanian). University Press, Cluj-Napoca (2005)

92. Kovacheva, R.: Zeros of sequences of partial sums and overconvergence. Serdica Math. J. **34**(2), 467–482 (2008)

93. Kurokawa, T.: On the Riesz and Bessel kernels as approximations of the identity. Sci. Rep. Kagoshima Univ. **30**, 31–45 (1981)

94. Lebedev, N.N.: (Translated from Russian by R.S. Silverman). Special Functions and Their Applications. Dover Publications, New York (1972)

95. Leviatan, D.: On the remainder in the approximation of functions by Bernstein-type Operators. J. Appox. Theory **2**, 400–409 (1969)
96. Lorentz, G.G.: Bernstein Polynomials, 2nd edn. Chelsea Publication, New York (1986)
97. Lorentz, G.G.: Approximation of Functions. Chelsea Publication, New York (1987)
98. Luh, W.: Univesal approximation properties of overconvergent power series on open sets. Anal. (Münich) **6**, 191–207 (1986)
99. Lupaş, A.: On Bernstein power series. Mathematica(Cluj) **8**(31), 287–296 (1966)
100. Lupaş, A.: Some properties of the linear positive operators, III. Revue d'Analyse Numer. Théor. Approx. **3**, 47–61 (1974)
101. Lupas, A.: Die Folge der Beta-Operatoren. Dissertation, Univ. Stuttgart, Stuttgart (1972)
102. Lupaş, L., Lupaş, A.: Polynomials of binomial type and approximation operators. Stud. Univ. "Babes-Bolyia", Math. **32**(4), 60–69 (1987)
103. Lupaş, A., Müller, M.: Approximationseigenschaften der Gammaoperatoren. Math. Zeitschr. **98**, 208–226 (1967)
104. Mahmudov, N.I.: Convergence properties and iterations for q-Stancu polynomials in compact disks. Comput. Math. Appl. **59**(12), 3763–3769 (2010)
105. Mahmudov, N.I.: Approximation properties of complex q-Szász-Mirakjan operators in compact disks. Comput. Math. Appl. **60**, 1784–1791 (2010)
106. Mahmudov, N.I., Kara, M.: Approximation theorems for generalized complex Kantorovich-type operators. J. Appl. Math. **2012**, Article ID 454579, 14 pages (2012). Doi:10.1155/2012/454579
107. Mason, J.C., Handscomb, D.C.: Chebyshev Polynomials. Chapman and Hall/CRC Press, Boca Raton-London-New York-Washington, D.C. (2003)
108. Mejlihzon, A.Z.: On the notion of monogenic quaternions (in Russian). Dokl. Akad. Nauk SSSR **59**, 431–434 (1948).
109. Meyer-König, W., Zeller, K.: Bernsteinsche Potenzreihen. Studia Math. **19**, 89–94 (1960)
110. Mocanu, P.T., Bulboacă, T., Sălăgean, Gr. St.: Geometric Function Theory of Univalent Functions, (in Romanian). Science Book's House, Cluj-Napoca (1999)
111. Mocică, G.: Problems of Special Functions (in Romanian). Edit. Didact. Pedag., Bucharest (1988)
112. Moisil, Gr.C.: Sur les quaternions monogènes. Bull. Sci. Math. (Paris) **LV**, 168–174 (1931)
113. Moldovan, G.: Discrete convolutions for functions of several variables and linear positive operators (Romanian). Stud. Univ. "Babes-Bolyai" Ser. Math. **19**(1), 51–57 (1974)
114. Mühlbach, G.: Verallgemeinerungen der Bernstein - und der Lagrangepolynome. Rev. Roumaine Math. Pures Appl. **15**(8), 1235–1252 (1970)
115. Muntean, I.: Functional Analysis (Romanian), vol. 1. "Babes-Bolyai" University Press, Faculty of Mathematics-Mechanics, Cluj (1973)
116. Obradović, M.: Simple sufficient conditions for univalence. Matem. Vesnik **49**, 241–244 (1997)
117. Ostrovska, S.: On the q-Bernstein polynomials and their iterates. Adv. Stud. Contemp. Math. **11**, 193–204 (2005)
118. Ostrovska, S.: q-Bernstein polynomials and their iterates. J. Approx. Theory **123**, 232–255 (2003)
119. Parvanov, P.P., Popov, B.D.: The limit case of Bernstein's operators with Jacobi weights. Math. Balkanica, N.S. **8**, 165–177 (1994)
120. Păltănea, R.: Sur une opérateur polynomial defini sur l'ensemble des fonctions intégrables. In: Itinerant Seminar on Functional Equations, Approximation and Convexity, (Cluj-Napoca), Preprint 83-2, Univ. "Babes-Bolyai", Cluj-Napoca, pp. 101–106 (1983)

121. Pǎltǎnea, R.: Inverse theorem for a polynomial operator. In: Itinerant Seminar on Functional Equations, Aproximation and Convexity (Cluj-Napoca), Preprint 85-6, University "Babes-Bolyai", Cluj-Napoca, pp. 149–152 (1985)

122. Pǎltǎnea, R.: Une classe générale d'operateur polynomiaux. Revue Anal. Numér. Théor. Approx. (Cluj) **17**(1), 49–52 (1988)

123. Phillips, G.M.: Bernstein polynomials based on the q-integers. Ann. Numer. Math. **4**, 511–518 (1997)

124. Radon, J.: Über die Randwertaufgaben beim logarithmischen Potential. Sitz.-Ber. Wien Akad. Wiss. Abt. IIa **128**, 1123–1167 (1919)

125. Rainville, E.D.: Special Functions. MacMillan, New York (1960)

126. Raşa, I.: On Soardi's Bernstein operators of second kind. In: Lupsa, L., Ivan, M. (eds.) Proceed. Conf. for Analysis, Functional Equations, Approximation and Convexity, pp. 264–271. Carpatica Press, Cluj-Napoca (1999)

127. Ruscheweyh, St., Salinas, L.C.: On the preservation of periodic monotonicity. Constr. Approx. **8**, 129–140 (1992)

128. Sauer, T.: The genuine Bernstein-Durrmeyer operator on a simplex. Results Math. **26**(1–2), 99–130 (1994)

129. Schoenberg, I.J.: On variation diminishing approximation methods. In: Langer, R. (ed.) On Numerical Approximation, pp. 249–274. University of Wisconsin Press, Madison (1959)

130. Schurer, F.: Linear positive operators in approximation theory. Math. Inst. Tech. Univ. Delft Report, (1962)

131. Sezer, S.: On approximation properties of the families of Flett and generalized Flett potentials. Int. J. Math. Anal. **3**(39), 1905–1915 (2009)

132. Soardi, P.: Bernstein polynomials and random walks on hypergroups. In: Herbert, H. (ed.) Proceedings of the 10th Oberwolfach Conference on Probability Measures on Groups, X, 1990, pp. 387–393. Plenum Publication, New York (1991)

133. Stancu, D.D.: Course of Numerical Analysis (in Romanian), Faculty of Mathematics and Mechanics. "Babes-Bolyai" University, Cluj (1977)

134. Stancu, D.D.: Approximation of functions by means of some new classes of linear polynomial operators. In: Colatz, L., Meinardus, G. (eds.) Proc. Conf. Math. Res. Inst. Oberwolfach Numerische Methoden der Approximationstheorie, 1971, pp. 187–203. Birkhäuser, Basel (1972)

135. Stepanets, A.I.: Classification and Approximation of Periodic Functions, Mathematics and Its Applications, vol. 333. Kluwer Academic, Dordrecht, Boston, London (1995)

136. Suetin, P.K.: Series of Faber Polynomials. Gordon and Breach, Amsterdam (1998)

137. Suffridge, T.: On univalent polynomials. J. Lond. Math. Soc. **44**, 496–504 (1969)

138. Szegö, G.: Orthogonal Polynomials, vol. **23**, 1939. American Mathematical Society, Colloquium Publications (1939)

139. Tonne, P.C.: On the convergence of Bernstein polynomials for some unbounded analytic functions. Proc. Am. Math. Soc. **22**, 1–6 (1969)

140. Uyhan, S.B., Gadjiev, A.D., Aliev, I.A.: On approximation properties of the parabolic potentials. Bull. Aust. Math. Soc. **74**, 449–460 (2006)

141. Vallée-Poussin, Ch. de la: Leçons sur L'Approximation des Fonctions d'une Variable Réelle. Chelsea Publication, Bronx, New York (1970)

142. Voronovskaja, E.V.: Determination de la forme asymptotique de l'approximation des fonctions par les polynômes de M. Bernstein (in Russian). C.R. (Dokl.) Acad. Sci. U.R.S.S. A **4**, 79–85 (1932)

143. Walsh, J.L.: Note on the coefficients of overconvergent power series. Bull. Am. Math. Soc. **48**(2), 163–166 (1942)

144. Walsh, J.L.: Interpolation and Approximation by Rational Functions in the Complex Domain, vol. XX, 5th edn. Amer. Math. Soc. Colloquium Publications, Providence, RI (1969)

145. Wang, H., Wu, X.Z.: Saturation of convergence for q-Bernstein polynomials in the case $q \geq 1$. J. Math. Anal. Appl. **337**, 744–750 (2008)
146. Wright, E.M.: The Bernstein approximation polynomials in the complex plane. J. Lond. Math. Soc. **5**, 265–269 (1930)
147. Wood, B.: On a generalized Bernstein polynomial of Jakimovski and Leviatan. Math. Zeitschr. **106**, 170–174 (1968)

Index

m-cusped hypocycloid, 82
m-leafed symmetric lemniscate, 82
q-Bernstein polynomials of quaternion
    variable, 99
q-Bernstein-Faber polynomials, 86
q-Bernstein-Kantorovich, 115
q-Durrmeyer, 115
q-Favard-Szász-Mirakjan operator, 83
q-Lorentz polynomials, 62
q-Lorentz-Faber polynomials, 115
q-Stancu polynomials, 73
q-Stancu-Chebyshev polynomials, 81
q-Stancu-Faber polynomials, 75
q-Szász-Kantorovich, 115

analytically continued Beta function, 21

Beatson kernel, 118
Beatson's kernels, 123
Bernstein's inequality, 3
Bernstein-Schurer polynomials, 13
Bessel function of the first kind, 182
Bessel type potential, 128
Beta operator of the first kind, 20
Bochner-Lebesgue integrable, 165
bounded turn, 121

Cauchy's formula, 2
Chebyshev and Legendre orthogonal
    polynomials, 172
Chebyshev orthogonal expansions, 164
Chebyshev polynomial of the first kind,
    124
circular lune, 83
close-to-convex, 122
complex Bessel type potential, 130
complex convolutions, 126

complex potentials, 128
convex, 122
convolution overconvergence phenomenon,
    viii
convolution polynomial, 122

de la Vallée-Poussin kernel, 118
differentiated Voronovskaja's formula, 5
Dini-Lipschitz condition, 166
Durrmeyer polynomials based on the
    Jacobi weights, 40

Euler's Beta function, 20

Faber coefficients, 5
Faber polynomial, 4
Faber series, 5
Fejér kernel, 118
Fejér-Korovkin kernel, 123
Flett potential, 129

Gegenbauer (or ultraspherical) polynomi-
    als, 183
generalized Jackson kernel, 118
generalized operators, 164
genuine Durrmeyer polynomials, 28

Hermite polynomials, 181
holomorphic, 99

Jackson kernel, 118
Jacobi polynomials, 183

Köebe function, 122
Korovkin kernel, 118

Printed in the United States
by Bookmasters

Printed in the United States
By Bookmasters